Modern Methods
in Horology

Also from Westphalia Press
westphaliapress.org

Modern Methods in Horology

A Book of Practical
Information for
Young Watchmakers

by Grant Hood

WESTPHALIA PRESS
An Imprint of Policy Studies Organization

Modern Methods in Horology: A Book of Practical Information
for Young Watchmakers
All Rights Reserved © 2018 by Policy Studies Organization

Westphalia Press
An imprint of Policy Studies Organization
1527 New Hampshire Ave., NW
Washington, D.C. 20036
info@ipsonet.org

ISBN-13: 978-1-63391-652-4
ISBN-10: 1-63391-652-9

Cover design by Jeffrey Barnes:
jbarnesbook.design

Daniel Gutierrez-Sandoval, Executive Director
PSO and Westphalia Press

Updated material and comments on this edition
can be found at the Westphalia Press website:
www.westphaliapress.org

MR. GRANT HOOD

MODERN METHODS

IN

HOROLOGY

By GRANT HOOD

A BOOK OF PRACTICAL INFORMATION FOR
YOUNG WATCHMAKERS

1913
BRADLEY POLYTECHNIC INSTITUTE
PEORIA, ILL.

PREFACE.

Knowing the difficulties that present themselves to the average watchmaker as he begins serving his apprenticeship and knowing how limited the supply of knowledge he is able to find and understand I have been prompted to write these pages, hoping the information may be such that it will encourage those that are discouraged, add renewed vigor to those who are ambitious and act as a warning to the ones that are inclined to be careless with their work.

My aim will be to make each subject as simple and clear as possible, adding illustrations in all cases where they are needed.

If the book is successful in helping my brother workmen and shall bring to them some new ideas that shall be beneficial or shall be the means of enabling them to do their work in an easier manner, the writer will feel that his labor has not been in vain and will be well pleased. GRANT HOOD.

Bradley Polytechnic Institute,
Peoria, Ill., April, 1903.

OLD AND NEW METHODS OF MEASURING TIME.

"What is time? The shadow on the dial, the striking of the clock, the running of the sand—day and night, summer and winter, months, years, centuries."

The measurement of time has been accomplished by various means for many centuries, in fact we are unable to trace its origin. Doubtless the first periods of division were those of day and night, caused by the revolution of the earth upon its axis, then upon closer observation it was noticed that the shadows of rocks, trees, mountains and hills grew shorter and shorter as the sun rose higher and higher until it reached a certain point when the shadows began to grow longer and longer toward the close of day or the setting of the sun. From these observations the first time piece of which we have any record was constructed, the sun-dial. The first record we have of a sun-dial is about 725 B. C., found in II Kings XX chapter, 11th verse—"And Isaiah, the prophet cried unto the Lord; and he brought the shadow ten degrees backward, by which it had gone down in the dial of Ahaz."

We read of some very large sun-dials constructed of massive stone masonry, among them the one at Benares. In olden times no other methods of telling time were known, even these on dark or cloudy days failed them.

The days have been divided into several different periods at various times, a decimal system was contemplated at one time, but the present system of twenty-four hours, each hour divided into sixty minutes and each minute into sixty seconds seems, perhaps on account of its familiarity to us, to be the very best possible one.

After the sun-dials, the clepsydra was invented. This was a very crude affair but had some points of merit the sun-dials did not have, viz., that of telling time when the sun did not shine and also in the evening after the sun had gone down.

7

MODERN METHODS IN HOROLOGY.

Several forms of clepsydrae or water clocks were constructed, some of them very ingenious, others very crude, even the best of them could not be depended upon as changes of temperature, different atmospheric pressures or the amount of water in them would greatly affect the flow. Some were vessels which contained water that escaped through small openings in the bottom, drop by drop, marks on the sides of the vessel indicating the hour; in another form the water trickled drop by drop into another vessel which contained a float, and this, as it slowly rose, would register the hour on a dial.

A very old method, used at Nepaul about the same time, was the floating of a copper vessel containing small holes in the bottom, on the surface of the water, the vessel was so constructed that in a certain period of time it would fill with water and sink. While doing so it would ring a gong, an attendant would empty the vessel and the process would be repeated indefinitely. The hour glass had about the same principle as the clepsydra, only sand was used instead of water. With all of these time pieces it would be necessary to have an attendant to watch them closely as when the water or sand ran out, no more time would be registered.

The sun-dials were the most common, and up to the middle of the nineteenth century were in common use in nearly all public places and many private families had them. A sun-dial will not tell the time accurately in any locality like a watch, but must be constructed for the latitude in which it is to be located. It will also be correct for all places either east or west on the same parallel of latitude, but the angle of the gnomen must be changed as we go north or south from that point.

To illustrate how far from correct the dials in actual use were, I show the photograph of one used in Western New York for nearly a century. On the dial is cast Lat. 40 degrees, 42 minutes. In looking up the latitude of the place I found it to be about 43 degrees, 20 minutes, and there for years this dial had been telling the time correctly.(?) Today the family who own this dial treasure it as one of their most valuable family relics. They watched me with suspicion even while I photographed it.

8

OLD AND NEW METHODS OF MEASURING TIME.

About the queerest method of telling the hours that has come to my notice was told me by one of the old settlers of New York. The country was new, lack of improvements were in evidence everywhere, clocks were only within the reach of the wealthy people, yet all must have some method of measuring time. Their novel method was this—

Sun Dial One Hundred Years Old.

on the window sill of the south window were a series of notches cut, the shadow from the casing as it was cast upon these notches indicated the hours.

Candles were in use at one time, they were made with ten rings in the tallow and contained the right amount that

9

it would take just one hour to burn from one ring to the next.

A lamp was similarly made. A graduated glass receptacle, which contained the oil, would denote by the marks the number of hours consumed in burning. These methods performed a double purpose of lighting the room as well as very incorrectly telling the time.

It will be impossible, within the scope of this article, to any more than allude to the many forms of clocks and watches that have been in use, which, being improved step by step, has brought us to the present state of perfection where it seems that there is no chance for improvement, yet if we could look ahead a hundred years, I fancy we would see as great an advancement over the present methods as the present methods are over those of a century ago.

Nothing has done more for the advancement of accurate time keepers than the railroads. Year by year the trains run faster and faster, a few seconds error may mean the loss of many lives, therefore the officials of the railroads require those responsible for the running of their trains to have the most perfect time pieces obtainable, and also require them to be inspected often and exercise every possible precaution to avoid error or accident.

Think of running the "Twentieth Century Limited" from New York to Chicago by the aid of an old verge watch without even a second hand, yet in its day, it was considered a marvelous time-keeper. The first clocks were constructed similar to the verge escapement. In place of the balance wheel was an upright piece with two arms upon which weights were hung. By moving them further out it would run slower, or faster when moved toward the center. This embraces the principle of the balances now in use, as is seen in the movement of the timing screws. As we carry the weight from the center of the balance wheel the watch will run slower and as we bring it nearer the center the watch will run faster. It follows that two balances of the same weight but of different diameters, the larger one will run more slowly than the smaller one as the mass of weight is farther from the center of rotation.

OLD AND NEW METHODS OF MEASURING TIME.

The lever escapement is in such general use today in the best class of watches where great accuracy and portability are required that we commonly speak of it as the leading escapement.

In later articles more detailed reference will be made to the duplex, cylinder and chronometer escapements.

This subject would not be complete without mention of one or two of the best clock escapements. The gravity escapement seems to be one of the best; we can add extra weight in the winter to force the hands on the dial of the tower through the snow and sleet or lessen the weight during the pleasant days of summer, yet the impulse to the pendulum remains the same, as the motive power only raises the weight which gives the impulse and the pendulum itself releases it, thus the amount of impulse remains constant regardless of the motive power.

The self-winding clock is a very ingenious piece of workmanship—it is so constructed that it winds each hour. A small electric motor winds up the thin spring enough to run the clock for the hour, at the end of that time a contact is formed which starts the motor, repeating the operation of the hour before.

Fig. 1—Naval Observatory, Washington, D. C.

TIME SERVICE OF TODAY.

There are but few people who realize the problems that present themselves in obtaining correct time. Never was more accurate time required than at the present, when railroads are spending thousands of dollars in order that they may be able to carry their passengers long distances in the shortest possible time. Improvements in rapid transportation and fine time pieces must go hand in hand. As better and more powerful engines are being built to make it possible to re- duce the time between different cities, even so must rapid advances be made in the manufacture of delicate time pieces which will enable them to run at such a rapid rate in safety. Thus far, the horologists have been able to keep slightly in advance, and the modern watches are truly wonderful for their fine workmanship and remarkable accuracy. In this, as in all branches of industry, as soon as a demand is created for a better article, someone is ready to supply it.

It was only a little over a century ago that watches and clocks then in use had but one hand, which denoted the hour only. In those days it was an improvement on methods then in use and seemed to satisfy the needs of that generation. As more accuracy was needed, another hand was added that di vided the hours into minutes. This, in time, was outgrown, and by the aid of a second hand, the hours and minutes were subdivided into seconds, just as accurately as before into hours and minutes. Today, by modern methods, it is easily possible to tell time to a thousandth part of a second.

A clock will stop, and we fail to wind our watch, it runs down; we ask a friend the time, or consult a jeweler's regu- lator, set our time piece correctly and think nothing further about it. Did you ever stop to think the jeweler must obtain his time from some source, and where he would go to get it? The object of this article and the following, will be to explain the methods adopted by the United States Government in ob- taining absolutely accurate time or as nearly so as it is possible to do with modern instruments of precision. It was my privilege recently to spend several days in the Naval Observatory at Washington, making a study of the methods now in use in taking observations and transmitting the time by telegraph throughout the United States.

In the outskirts of Washington, somewhat isolated from the rest of the city, is situated the new Naval Observatory. Perhaps less is known of what takes place here by the citizens

Fig. 2—Lieutenant-Commander Hayden at His Desk.

of Washington than of any of the Government buildings. It is necessary to have the building as remote as possible from the street railways and the rumble of the city, as their vibrations would interfere with the delicate instruments in use. When one has passed through the narrow lane leading to the grounds and climbed the hill, a beautiful sight presents itself. (Fig. 1.) The beautiful white stone buildings of the Naval Observatory, with their large circular domes surmounting them, is an inspiring sight. Our Government here, as is the case with all of its buildings, has planned them in a most substantial manner, embracing beautiful architecture and pleasant surroundings. From this place comes our time. In other words, this is "Uncle Sam's time factory."

As I entered the building, the guide took me at once to the office of the Lieutenant-Commander, Edward Everett Hayden, whom I found seated at his desk (Fig. 2), busily engaged, yet he had ample time to give me a hearty greeting,

Fig. 3—The Transit Instrument for Observations.

and made me feel at home immediately. To him especially, and to others connected with the Observatory, I am greatly indebted for many courtesies and much valuable information. It would be impossible in two articles to explain the time system thoroughly, yet I trust it may give a much better idea of an important system that to most people is entirely unknown. It is commonly understood that our time is taken from the sun as it passes the meridian at noon. This is not the case, as the sun passes the meridian but four days in the year exactly at noon. The observations are taken from certain fixed stars by the transit instrument shown in Fig. 3. The utmost accuracy must be used in making and setting up a transit. It points due north and south, and can be placed in any position from the vertical to the horizontal, but moves only in one plane. There are many fine adjustments to test its accuracy and errors. The building that contains the transit circle is made entirely of iron, a sheet-iron covering

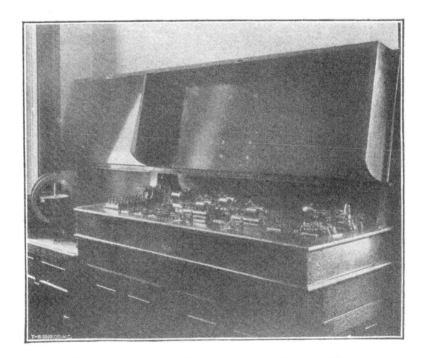

Fig. 4—Instrument Used for Sending Out the Time.

on the outside and the inside lined with the same material, having an air space between. There is no way of heating the room, as it is necessary to have the temperature inside as nearly that out-of-doors as is possible, to produce the best results. The transit is mounted on massive stone piers, which extend many feet below the surface of the ground, in order to get a perfectly solid foundation. The floor of this building does not touch the stone base at any point. In other words, the building is only a protection from the sun and storms, the instrument itself standing on its own foundation.

The eye piece of this transit contains eleven vertical lines or spider webs, a group of three on the left, five in the center and three on the right, a horizontal one crossing all of them at the center. At the Observatory are tables showing the exact moment the fixed stars will pass the meridian, and their position. By careful graduated circles the transit instrument is set at the proper angle and the person making the observation assumes a lying position in the adjustable chair shown and patiently waits for the appearance of the star. In one hand he has an electric button, which is connected with the chronograph, shown at the right in Fig. 5. This consists of a cylinder, around which a sheet of paper is placed, the cylinder making one revolution every minute, and is also connected with a standard clock. Each vibration of the pendulum is recorded on the paper by a fountain pen. As the image of the star reaches the first vertical line, the electric button is pressed and the exact time recorded on the chronograph, and in a like manner when it passes each of the eleven lines. The average is then taken of them all, which gives the exact time when the star passes the center line, or meridian. By taking several observations in one night, it is possible to get the time to a very small fraction of a second. From the record on the chronograph, the exact siderial time is found, and from this is computed a local standard time. Each day at noon, the time is sent out to every city and town in the United States, east of the Rocky Mountains, by telegraph. The time balls are dropped in the principal cities along the sea coast and at the Navy Yards. Clocks are set to the second, and now bells are rung on many telephones, all by the electrical current sent out by the standard clock at Washington. To stand near this clock and see its pendulum vibrate to and fro, measuring the seconds of time so accurately, and to think that its vibrations can be heard in all cities throughout this vast land, seems indeed one of the great achievements of the present century.

Figure 4 shows the instruments used for sending out the time over six thousand miles or more of wire throughout the United States east of the Rocky Mountains. More will be said about these instruments in the following article.

17

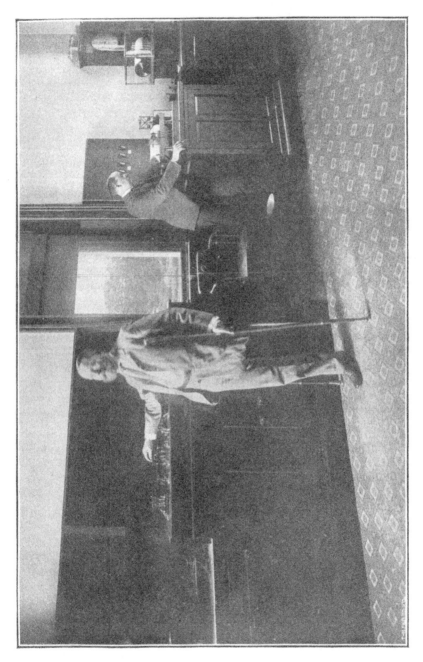

Fig. 5—Lieutenant Hayden Stands Watching the Standard Clock, Ready to Transmit the Correct Time.

TIME SERVICE OF TODAY.

Figure 5 shows the various telegraph instruments of the Western Union and Postal Telegraph Companies. The eight point relay which sends out the current that synchronizes the clocks, drops the time balls, etc. Lieutenant Hayden stands before the instruments with his hand on the lever, watching the second hand of the Standard Clock, which has been corrected to the second only a few moments before. At the exact moment the lever is thrown down, the pulse beats of the clock can be heard throughout the United States. The clock which sends out this current is corrected daily a few minutes before noon. It matters but little how slow or fast it is throughout the day or night, but for the five minutes before 12 o'clock it is supposed to be absolutely correct. Its corrections are made in a peculiar manner. Should it be slightly slow, the pendulum is quickened by a touch of the hand, or should it be too fast it is retarded in the same manner until its time is corrected to a small fraction of a second.

There are many things that will vary the rate of even the most carefully constructed time pieces, the changes of temperature, atmospheric pressure, thickening of the oil, etc. In order to overcome these difficulties, the Government has recently been making some careful tests and experiments. A clock vault has been built underground and equipped with costly apparatus that will keep it at a constant temperature. The outer walls of the vault are made of brick; an air space of about ten inches is left between this wall and the inside ones, which are of wood, covered with asbestos. In this air space are placed several coils of pipes through which the hot water circulates that keeps the room at a constant temperature of about 80 degrees F. The water is heated by a small gas heater, which is automatically controlled by a very sensitive thermostat located inside of the vault and operated by electricity. The temperature will vary only about one-fourth to one-half of a degree. This vault is a success as far as constant temperature is concerned, but as the barometric pressure of the atmosphere affects the time of the clocks as greatly as the variable temperatures, this has yet to be overcome, and now experiments are being made, enclosing the clocks in air tight cases and exhausting the air, but such cases are difficult to construct that maintain a vacuum.

Fig. 6.—Empire State Express Running 70 Miles an Hour.

20

Thus far it has been only partially successful. The only way of its success being assured seems to be in making a large glass globe similar to the ones used on air pumps, and exhausting the air.

This shows to what extent our Government will expend time and money to perfect a system that is of vital importance to every citizen in the land.

The time of all places having the same longitude will be the same, those places east of that longitude the time will be later, and those west from it will be earlier. This is caused by the rotation of the earth upon its axis once every twenty-four hours. Each city, then, must have its local time, while those cities eastward or westward must have local time faster or slower in exact proportion to the change of longitude, there being one hour for each fifteen degrees; for example, if we go eastward, the time will be an hour faster for every fifteen degrees we travel, and likewise if we go westward, the time will be one hour slower for every fifteen degrees we travel.

For a practical illustration, let us suppose two people start from Greenwich which is located in longitude 0° 0', each intend to go around the world, one in an easterly and the other in a westerly direction; their watches are set alike, Greenwich time; as the former compares his time with that of the cities he passes through, he finds their time faster and faster the further east he goes until when he reaches the 180° E. longitude, he finds the time exactly twelve hours ahead of his Greenwich time. In like manner the one who travels westward finds the time constantly growing slower than his Greenwich time, so by the time he reaches the 180° W. longitude, the time is just twelve hours slower than his. One woud reach the 180th meridian on Thursday and the other on Wednesday, yet their watches are still the same time. For this reason the captains of vessels either drop a day or gain one as they cross this meridian. At sea all longitude is reckoned from Greenwich and all ships' chronometers (each vessel having two or three) are set to Greenwich time. I noticed while at the Naval Observatory, where all the chronometers of the navy are rated for many weeks before sending them to the men of war or cruisers, that the

21

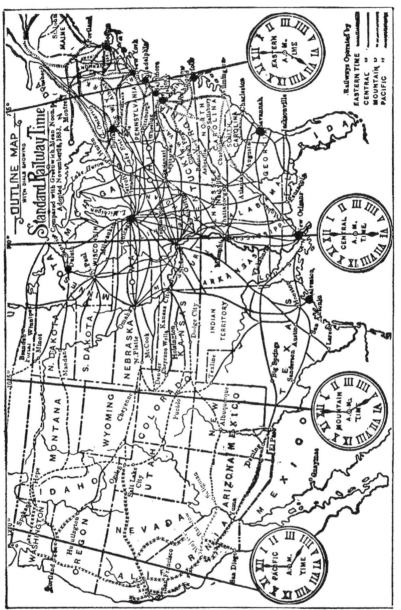

Fig. 7—Map of Standard Railway Time System.

time at which they were set was a trifle more than five hours different from the local time of Washington, there being a difference of about 77° of longitude between Greenwich and Washington.

As each city has its local time, which differs from other

Fig. 8—Time Ball on Top of State, War and Navy Building, Washington, D. C.

cities in proportion to their difference in longitude, it would make it very confusing to run trains on the railroads according to such local times without complicating matters, and it would be very hard to avoid serious accidents.

23

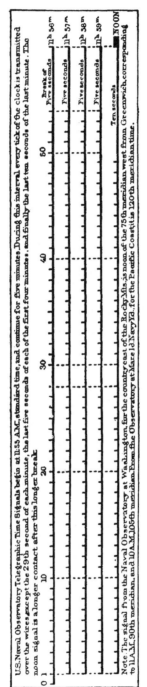

U.S. Naval Observatory Telegraphic Time Signals begin at 11.55 A.M. standard time, and continue for five minutes. During this interval every tick of the clock is transmitted over the wires except the 29th second of each minute, the last five seconds of the first four minutes, and finally the last ten seconds of the last minute. The noon signal is a longer contact after this longer break.

Note. The signal from the Naval Observatory at Washington, for the country east of the Rocky Mts., is noon of the 75th meridian west from Greenwich, corresponding to 11 A.M. 90th meridian, and 10 A.M. 105th meridian. From the Observatory at Mare Id. Navy Yd., for the Pacific Coast, it is noon of the 120th meridian time.

Fig. 9.—Chart Showing How the Time Is Sent Out by the Standard Clock in Washington.

Figure 6 shows the Empire State Express going at a rate of seventy miles an hour, which could not be done without some system which would prevent accidents.

To overcome the difficulties that were common when the local times were in use, a standard time was adopted in 1883, which has overcome all of the then existing troubles and makes it possible to run trains as fast as any motive

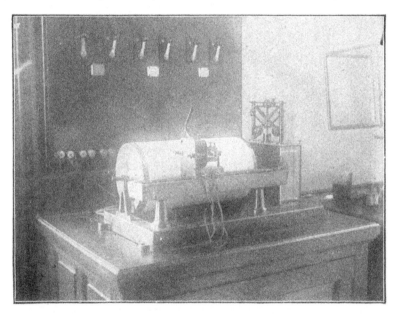

Fig. 10—The Chronograph Recorder.

power in existence can pull them. Each year the speed of famous trains is increased and as rapidly as this is done, the timepieces must be made to perform more closely, and the railroads require their men to carry a higher grade of movements. A watch that five or ten years ago was considered a perfectly satisfactory timepiece would not be allowed in the pocket of an ergineer today.

Our standard time is divided into sections of 15° each, all places of each section having the same time. In the United States the meridians adopted are those of 75°, 90°, 105°, and 120° west of Greenwich, these being 15° apart, or exactly one *twenty-fourth* part of the earth's circumference. It is easily seen that each 15° will represent exactly one hour of time. The 75° meridian is Eastern time, the 90° Central time, the 105° Mountain time and the 120° Pacific time.

The time is the same at all points situated between meridians either 7½° east or west of the ones above mentioned.

At Buffalo, N. Y., for example, when the time changes from Eastern to Central, you may enter the city on a train

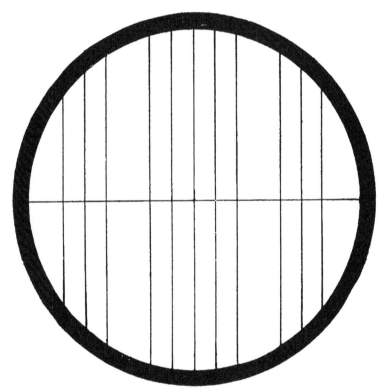

Fig. 11—Eye-Piece of Transit Instrument for Observation.

from the east and immediately leave for the west, and the time will be just one hour slower; this time will remain in use until we travel 15° farther, when our time will be one hour slower, etc., until we reach the Pacific coast.

When it is twelve o'clock or noon at Washington (Eastern time), it will be eleven o'clock Central time, ten o'clock Mountain time and nine o'clock Pacific time.

Figure 7. The railroads do not change their time exactly on the meridian but at some prominent station near that point. The map shown gives a fair idea of the different times as now in use by the railroads. The dark spots locate the cities where the government has time balls located. Most of these places are near the sea coast where the time balls may be seen from the vessels at sea, in order that the officers may compare their chronometers by watching the time balls as they drop at noon. Figure 8 shows the one at the top of the State, War and Navy Building in Washington, D. C. Five minutes before noon a signal is sent out from the Observatory and the ball is raised to the top of the pole and held in place by a catch; during the last ten seconds before noon a switch is thrown connecting the time ball with the standard clock, which at the exact second of noon sends out the current that releases the catch and drops all of the time balls in the sixteen cities at the same instant. They drop into a large circular receptacle, slightly larger than the ball; this forms an air cushion which lessens the concussion of the fall. The diameter of these balls is about three feet, and they fall from fifteen to twenty feet.

Figure 9 is a reproduction of a chart from the observatory showing how the time is sent out by the standard clock in Washington for the five minutes before noon each day and recorded on the chronograph (Fig. 10). The twenty-ninth second is omitted and the last five seconds of the 55th, 56th, 57th and 58th minutes are omitted, but in the 59th minute the last ten seconds are omitted. During this long break, the time balls are switched into the circuit.

Any one who may be at a telegraph office during this time may know that the *minute begins* after the five second break, and the half minute or 30th second is after the single break.

In the large cities the watchmaker can have a "ticker" placed on his bench connected with a standard clock of the Telegraph Company which tells the minutes in much the same manner as just described.

The Telegraph Companies also furnish clocks which are connected by electric wires to a master clock which sends out a current each hour, correcting or synchronizing them; all clocks on that circuit being corrected at the same instant.

Figure 11 is a drawing showing the construction of the eye-piece of the transit instrument described in the last article. As the star passes each line a record is made on the chronograph (Fig. 10).

The beats of the pendulum of the clock are also recorded so that by comparison the exact moment the star passes the center "spider web" can be easily computed.

Modern methods of obtaining accurate time are as far in advance of those of a century ago as the modern, finely adjusted watch is ahead of the verge of the same period, which is now a curiosity and only found in museums or in collections of antique watches.

Will the century ahead of us show as great advancement?

IRON AND STEEL.

In beginning a series of articles on modern methods used in horology, there is nothing that interests the watchmaker or that should be better understood than the proper working of iron and steel, yet there is no department of his work that seems to be more neglected. A spring will break or lose its elasticity, a graver will fail to hold its point, or it is impossible to obtain a high polish on another piece of steel. These are common effects, but their causes are not known by the ordinary workman. He knows that these faults exist, but he does not know how to remedy them.

The idea occurred to me a few years ago, that by the aid of the microscope, we might be able to study the grain of the steel we are using, and possibly be able to remove existing trouble, and with that end in view, I began a series of experiments which proved of great value in my work, and I hope may be as helpful to my readers. The illustrations in this article are from photographs taken with the microscope and show various magnifications from 24 to 150 diameters.

IRON.

Iron is an elementary body, and is one of the most common and useful metals. In some form, it is used in nearly all branches of industry. In its ordinary form, it is but of little use in horology, but when converted into steel, it is used in making the finest tools and most delicate parts of the time pieces used at the present day. Iron as in common use, is known under three names, viz., cast iron, wrought or malleable iron and steel. The watchmaker has but little to do with the first two, while the last, steel, he is dependent upon for his various tools and the construction and repairing of delicate time pieces, and for the making of many of their parts.

I do not wish to make these articles too technical, but in order to thoroughly understand the processes necessary for the best working methods of handling steel, we must have an idea of its composition. But very few of our best workmen realize the slight differences that exist in a composition of iron and steel. Their main difference being the amount of carbon they contain.

Cast iron is the cheapest and most common kind, and for some purposes is far superior to any other form. It is very brittle and a broken surface is coarse and crystallized. Cast iron cannot be bent, is not malleable, and therefore is not of use to us except in the manufacture of tower clocks or in constructing large machinery. It contains about 2 per cent. of carbon.

WROUGHT OR MALLEABLE IRON.

Wrought or malleable iron, contains the least carbon, its amount being about 5-10 of 1 per cent. It can be rolled into sheets, drawn into wire or forged into any desirable form. It can be made from cast iron by removing some of its carbon.

STEEL.

The third form, steel, is one of greatest interest to us, and to a study of this, we will direct our attention. We know that if we heat a piece of steel to a red heat, and plunge it into water, it becomes very hard and brittle, but why it hardens no one seems to be able to explain satisfactorily. There are many theories, but these are of no great interest to us; we wish to know how to handle it to the best advantage. There are three forms of steel in use; first, natural steel; second, shear steel; third, cast steel.

Natural steel is made from wrought iron by heating for several days with charcoal. The carbon in the charcoal unites with the iron, converting it into steel, but that made by this method is far inferior to cast steel.

Shear steel is made by binding several bars of steel together, and forging and welding them into a solid piece, this process being repeated several times. Upon being ground and polished, steel made by this method shows by the streaks on the surface where the different bars have been welded together, and therefore is but of little use to us. Fortunately, however, none of this class of steel is in use at the present date.

Cast steel is used exclusively for the manufacture of fine

tools and delicate articles. It is always used whenever a superior grade is required, yet even in cast steel, we find many different grades, in fact if you state to the manufacturer the purpose for which you desire its use, he will gladly select the grade most suitable for its purpose. This

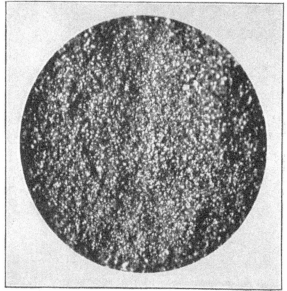

Fig. 1—Steel hardened at white heat (badly burnt).

is quite important, as the kind that would make the best large tools might make the very poorest pivot drills or delicate pieces of watches.

We must become thoroughly acquainted with the steel that we are using, and when once familiar with it, should not change for other brands as each kind requires its own special way of handling. A degree of heat that would nicely harden a low grade of steel applied to that of a higher grade, would burn it so that it would be practically ruined. From this we learn the following: The higher the grade of steel, the lower the temperature at which it hardens; and the lower the grade, the higher the temperature required in hardening. As we use only that of the very highest grade in our work, and our articles are the most delicate, we must be doubly careful about heating in order to prevent it from "burning."

31

HARDENING AND TEMPERING.

These terms are often misused, as we often hear one speak of tempering a piece of steel, when in reality it has only been hardened. We harden a piece of steel by heating

Fig. 2—Steel hardened at bright red (slightly burnt).

to a dull cherry red and plunging it into water, oil or any substance that will quickly cool it.

Steel hardened in mercury, nitric acid or cyanide of potassium will be very hard and brittle while that hardened in oil, tallow or bees-wax will be quite hard and very tough.

One of the most important things to keep in mind in heating the article to be hardened is to heat all parts *evenly* —to illustrate: We wish to harden a piece of small round wire and hold it in the blaze of our lamp; one side of that wire will be heated to a bright red and the other to a dull red. We know that heat expands all metals and it is clearly seen that the side of the wire that is heated the hottest will be the longest, by cooling quickly; the molecules have not time to resume their ordinary form but become crystallized in that form, the wire being the longest on the side heated the hottest. To overcome this trouble the wire should be constantly rolled in the flame in order to heat all sides equally, and it is always best to plunge the piece into the oil or water lengthwise and do it very quickly. The oxygen of the air unites with the steel on its surface when heated and forms an oxide which in some cases is hard to remove. It is possible to harden a piece of polished steel without affecting its polish to any great extent. We have just learned the

Fig. 3—Steel hardened at lowest possible heat (dull cherry red).

cause of the oxide, and if we can heat the piece without its coming in contact with the air, we can accomplish the desired result; this may be done by heating in a copper tube or the bowl of a clay pipe and filling the tube or pipe with carefully dried animal charcoal; then heating to a bright cherry red and throwing the contents of the tube into the water the charcoal burns the oxygen and prevents it from acting on the bright surface of the steel. Delicate articles are often hardened in this manner after which they can be tempered without being obliged to polish them.

Let us now carefully examine the photographs here reproduced. Figs. 1, 2 and 3 were made from the same piece of steel, a rod 5 millimeters in diameter, in order to show the effect of overheating or "burning"; the end of the rod was heated to a white heat; a short distance from the end of it was a bright red, and still further back, it was not hardened at all. We have in this short piece of rod all stages,

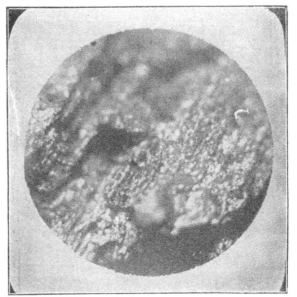

Fig. 4—Wrought iron showing grain.

from the badly burned to the most perfectly hardened steel possible.

The rod was easily broken at the end and showed a very coarse crystallized surface as shown in Fig. 1. The grain resembled very much that of cast iron and its brittleness was much the same; this represents the "burnt" steel which we wish to avoid and was hardened at a *white* heat.

Fig. 2 was broken off at the point where heated to a bright red; we still have a coarse grain, but not so pronounced as in Fig. 1. It was not as brittle and required much more effort to break it.

Fig. 3 was broken off at the point where it was hardened at its lowest temperature or barely a red heat. This did not break so easily; in fact, it took several blows with a heavy hammer to break it. We notice a great change in the fracture; the surface is very smooth, the grain beautiful and fine, and the coarse crystals have entirely disappeared. Is it any

wonder that delicate springs often break or that edge tools
will not remain sharp when we can see so clearly the effect
of overheating in hardening?

In hardening all classes of steel, we should heat it to the

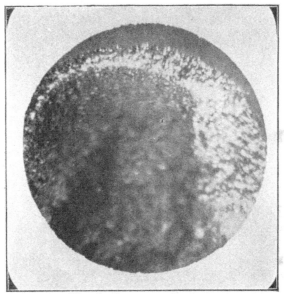

Fig. 5—Iron case hardened.

lowest possible degree, a dull red being enough for most high
grade steel; if overheated it is liable to check or crack.

TEMPERING.

After hardening there is a great strain between the
molecules; pieces have been known to fracture in many
pieces days after hardening. For this reason and on account
of its extreme brittleness, it is necessary to temper the
steel, each piece being drawn until it is of the proper hardness for the purpose required.

The common method is to polish the surface, then by
heating carefully, watch the oxide as it forms upon the
surface. Each color denotes a certain hardness, as follows:

1. Very pale straw.
2. Straw.
3. Dark straw.
4. Brown yellow.
5. Purple.

6. Light purple.
7. Dark purple.
8. Dark blue.
9. Light blue.
10. Pale blue and green.

The first two are too hard to file and denote the right temper for gravers, cutters, etc.

Three and four are about right for dies and taps.

Five to eight will do nicely for staffs, pinions, springs, etc.

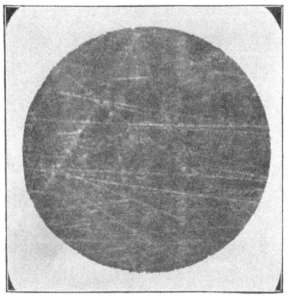

Fig. 6—Steel highly polished.

To obtain a nice even color, the *utmost cleanliness* must be observed, even a finger mark being sufficient to prevent a color that otherwise might have been perfect.

The finer and brighter the polish before tempering, the better and more even the color will be.

If we temper an article without its coming in contact with the oxygen of the air, no oxide will be formed; we take advantage of this method by tempering in oil, which for some purposes is the most satisfactory method known.

By placing a piece of hardened steel in a small cup of lard oil sufficient to cover it completely and heating slowly over the Bunsen burner or alcohol lamp, we soon notice the surface of the oil beginning to smoke; when this takes place our steel will be tempered to a light straw; when the oil smokes densely, a dark straw; when the surface ignites, a purple; when the oil burns, it will correspond to a blue; and when the oil burns up, the steel will have a spring temper.

There is no better way of tempering case springs and the various small springs in the movements than by placing them (after hardening) in a common iron spoon, covering with lard or bees-wax, and heating until the oil or wax ignites and burns off, and repeating the operation in some cases two or three times.

It is possible to temper more evenly in oil than in the open flame as the oil surrounds the piece and heats it evenly on all sides, while in the open flame, one side is liable to become more heated than the other.

ANNEALING.

Should we fail to heat an article quite hot enough to harden and plunge into water, we will be surprised to find it softer than before. This we call water annealing, and is a quick and useful method of making the steel soft and easily workable.

Another method is to heat to a red heat and let it cool very slowly while being covered with some substance that is a non-conductor of heat, such as ashes, lime, etc.

CASE-HARDENING.

In some cases it is useful to know how to convert the surface of iron into steel, which is called case-hardening. Wrought or malleable iron has a grain similar to that of a piece of wood. Fig. 4 shows the fracture of this metal, the surface being so uneven it was impossible to get all parts in focus while making the photograph. Should the surface of this iron be covered with the yellow prussiate of potash and heated carefully, or should it be enclosed in an iron box filled with pieces of leather, horn or similar substances and heated for several hours, the surface would be converted into steel, and after being hardened, would present the appearance of Fig. 5, which is an ordinary small wire nail case hardened. This represents only one side of the nail, the light portion being the part converted into steel.

The watchmaker is perhaps as greatly interested in knowing how to obtain a fine polish on his finished work as in anything connected with the working of steel. In a later article, this will be dealt with very fully.

MODERN METHODS IN HOROLOGY.

The utmost care must be exercised in the different stages not to have any of the coarser grinding materials enter in any way into the finer ones; this one point has been the most difficult one in my experience to impress upon the minds of those just beginning.

If the carpenter has a board in the rough and he wishes to prepare it for its final polish, he just planes it, then sand-papers it with coarse sand paper and lastly with the very finest, and it gradually assumes a perfectly smooth surface; even so must the watchmaker obtain a satisfactory polish. First, we grind the surface on a lap or iron grinder with oil stone powder and oil until it assumes a flat smooth surface. This must now be thoroughly cleaned to remove all traces of the oil stone powder, as should a particle remain it will prevent the finer material from doing its work.

After being thoroughly ground, we should repeat the same operation with either crocus and oil or coarse diamontine and oil, after which it should be as thoroughly cleaned as before, and the final polish be given on a lap or polisher made of equal parts, of tin and zinc. The material used in the final polish should be only the very best quality of diamontine that can be obtained. This is mixed to a thick paste with oil and only a small amount used on the lap, the polish not being complete until the oil is nearly dry on the lap. There is a "knack" about polishing that can only be obtained by experience, therefore the novice must not be discouraged if he does not succeed the first time, but try again with renewed vigor, and success will surely crown his efforts.

The eye is very easily deceived. If we can finish a flat surface so that it looks flat, and appears to be thoroughly polished, we may be satisfied, for Fig. 6 shows us that even with all of our pains our work is very imperfect. This represents the surface of a flat piece of steel that to the unaided eye was well polished, yet when viewed under the compound microscope, presented the appearance shown in the photograph. In future articles, many other photographs will be shown, even more striking than those contained in this article.

WHEELS AND PINIONS.

The use of a train of wheels, containing teeth, as a means of transmission of power dates back many centuries. It was in use by the Egyptians and the Romans, although their wheels were poorly constructed, badly spaced and the teeth of very irregular form; in fact, the workmanship was so faulty that it was impossible to transmit power without great loss; the motive power had to be great and the consequent wear would be excessive. Less than a century ago the teeth in the wheels were rounded up by hand; if some of the verge watches of that period are carefully examined the teeth of the wheels will show very faulty workmanship in comparison with the high degree of perfection attained at the present day.

It will not be the purpose of these articles to show how the epicycloidal curves are generated or formed, as that would be of little use to the watchmaker, but to show him how to determine the correct sizes of any lost wheels or pinions, and also to be able to quickly tell the number of teeth or leaves in them.

There are two forms of trains in common use, Simple and Compound. In a simple train, the first wheel transmits its power to the one following it, this in turn transmits the power to the one following it and so on throughout the train, the circumferences of *all of the wheels travel at the same velocity* and each wheel turns in the opposite direction to the one before it or the one following it. In finding the revolutions of the last wheel to one of the first wheel in such a train, it is only necessary to compare the sizes or number of teeth of the former to those of the latter, the intermediate wheels only acting as a means of transmission of power. If the last wheel had 20 teeth, and the first one had 40 teeth, then the last one would make two revolutions to one of the first. Figure 1 illustrates such a simple train.

While A makes one revolution, B would make two, but

39

the speed of the circumference of each would be the same, as would also that of each of the intermediate wheels.

Should the sizes of A and B be the same, then the power applied to the surface of A would be exerted on the surface

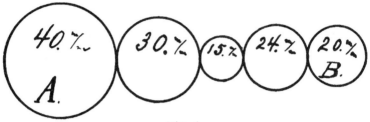

Fig. 1.

of B less that wasted by friction, a force ever present and one the watchmaker tries in so many ways to overcome.

In a compound train, the kind used in watches and clocks, the first wheel transmits its power to a pinion following it; upon this pinion is staked a wheel, which in turn transmits its power to a pinion following it, and in like

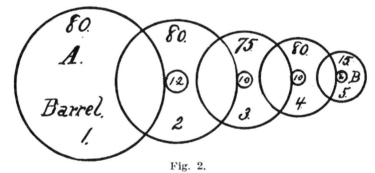

Fig. 2.

manner this power is transmitted to the end of the train. Each wheel revolves more rapidly than the one before it, and the power grows less and less until a very small amount of power is required to counteract all of that exerted at the beginning of the train; "while we gain in speed, we lose in power," that rule of physics is very forcibly shown here.

We see a man on a derrick turning a crank which turns a small wheel, this in turn engages a larger one and so on throughout the train of wheels, he is turning the crank rapidly and by the aid of these wheels he is exerting tremendous power and is raising a weight many times his own; he is doing it very slowly, what is gained in power, is lost in speed. The same train is used here as in watches and clocks, only in this case the power is applied at the *opposite* end of the train.

In an ordinary watch, while the barrel makes one revolution the escape wheel and pinion will make 4,000 revolutions.

Figure 2 shows a compound train as above, containing teeth in the wheels and leaves in the pinions as follows:

Barrel, 80 teeth.	Center pinion, 12 leaves.
Center wheel, 80 teeth.	Third pinion, 10 leaves.
Third wheel, 75 teeth.	Fourth pinion, 10 leaves.
Fourth wheel, 80 teeth.	Escape pinion, 8 leaves.

To find the revolutions of the last to one of the first in any train, multiply the teeth in all of the wheels (working

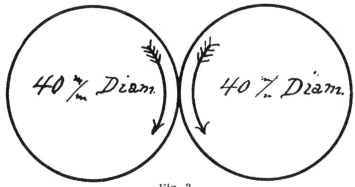

Fig. 3.

into pinions) together and divide that product by that of all of the leaves of the pinions working into these wheels, the quotient will be the revolutions of the last to one of the first, as:

$$\frac{80 \times 80 \times 75 \times 80}{12 \times 10 \times 10 \times 8} = \frac{\overset{2}{8} \times 25 \times 10 \times 80 \times 75 \times 80}{12 \times 10 \times 10 \times 8} = 4000$$

41

In order to make this as clear to the novice as possible, we will consider the wheels *without* teeth first and think of the surfaces of the wheels and pinions as perfectly smooth and turning each other by friction. This, of course, could not be in actual practice, as their surfaces would be inclined to slip and the friction, the one thing we must avoid, would be excessive, but as we are able to show their geometrical diameters more clearly in this manner, we will do so.

In Figure 3 we have two wheels of the same diameter; if we should turn A exactly one revolution and the surface

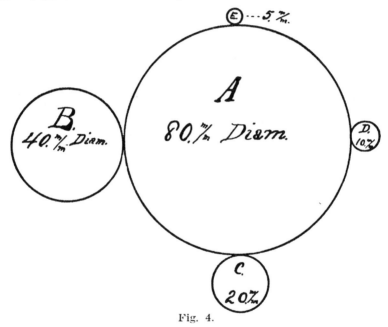

Fig. 4.

of B were in contact with that of A, it can be clearly seen that B would also make one revolution because the circumferences of each are equal. From this we learn that wheels having the same diameters working together make the same number of revolutions and turn in opposite directions.

In Figure 4 we have one large wheel and four smaller ones turned by it. They are, respectively, one-half, one-fourth, one-eighth and one-sixteenth the diameter of the large wheel.

WHEELS AND PINIONS.

The wheel A has a diameter of 80 m-m, B is one-half that of A, or 40 m-m; if the diameter of B is one-half that of A, then the circumference also must be one-half that of A. The speed of their circumferences must be the same as they are turning by friction; this being true, we can easily see that B must make two revolutions to one of A. The diameter of C is *one-fourth* that of A, therefore it must make four revolutions while A makes one. D being *one-eighth* that of A, it must make eight revolutions while A makes one, and E being one-sixteenth that of A, must make sixteen revolutions to one of A.

These points should be thoroughly understood before we attempt to compute trains, as unless the first principles are well understood, it will be quite difficult to comprehend the more difficult problem that will follow:

In Figure 4 should the power be applied to E, it would have to make sixteen revolutions before A would make one. This illustrates the power of man as applied to the crank, raising the heavy weight, while the power as applied to a watch train would be that of A revolving E sixteen times to one of itself.

It is not practicable to convey the power from the barrel to the escapement with only two or three wheels, as it is necessary that the center wheel should make one revolution

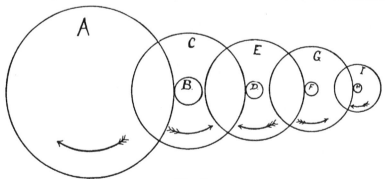

Fig. 5.

each hour and the fourth wheel should revolve once each minute, and both must turn in the same direction. This can only be accomplished by using a compound train, which is illustrated in Figure 5.

43

A represents the main wheel or barrel which contains the spring or motive power of the watch. The teeth of A engage the leaves of the pinion B, turning it in the opposite direction; upon B is staked the wheel C, which must turn with it. Wheel C engages pinion D, upon which is staked wheel E. This in turn engages pinion F, which is the arbor upon which wheel G revolves. This wheel engages the last pinion H.

The larger one of a pair working together is commonly called a wheel, and the smaller one a pinion.

It will be noticed in all compound trains that each alternate wheel and pinion revolve in the same direction.

The projections on the edge of a wheel are called teeth and work into the spaces of the pinion; the projections on the surface of the pinion are called leaves, and work into the spaces between the teeth of the wheel.

The pointed part of the tooth projecting beyond the pitch circle or primitive circumference is called the *addenda*, that of the pinion the *rounding*.

Great care is necessary in forming the shapes of the teeth and leaves, as a smooth, easily working train can not be produced otherwise. Various forms of teeth are employed according to the work to be performed. Those that would work nicely in some cases would be very faulty in others, so that it is necessary to have different methods of finding the sizes according to the uses required.

The rules following will enable the reader to figure out the correct sizes of all wheels and pinions according to methods now in use in most of the leading watch factories.

In speaking of the train of a watch, we mean a system of wheels and pinions used to transmit the power from the main spring or barrel to the escapement, which regulates the speed of the wheels, and at the same time makes it possible to accurately register the time by the revolutions of the hands.

Each of the wheels and pinions forming the train have a technical name, which should be well understood.

The barrel, containing the motive power or main spring, is called the *first* wheel. We do not often call it by this name, but speak of it as the "barrel."

WHEELS AND PINIONS.

The teeth of the barrel act upon the leaves of the pinion usually found in the center of the watch, called the "center pinion." There are some cases where this pinion is located elsewhere, as with movements with a sweep second hand.

The *center* or *second* wheel is staked upon the center pinion and its teeth engage the leaves of the *third* pinion, upon which is staked the third wheel. This in turn acts upon the *fourth* pinion carrying the *fourth* wheel. The fourth wheel engages the fifth pinion or, as we commonly call it, the escape pinion, and upon this pinion is staked the escape wheel or fifth wheel.

It will be seen that we begin numbering the wheels at the barrel, and count towards the escape wheel, as first, second, third, fourth and fifth or escape. The pinions, it will be noticed, take their names from the wheels which are staked upon them, as center wheel and pinion, third wheel and pinion; while in reality the third pinion is only the second one, it would be very confusing to number them in any other manner.

A very important thing for us to learn and understand thoroughly is, that in finding the sizes of wheels or pinions, we must always consider them in *pairs;* the pair will be the *wheel and pinion working together.* The train of a watch consists of the following pairs: Barrel and center pinion; Center wheel and third pinion; Third wheel and fourth pinion; Fourth wheel and escape pinion.

We have learned that the revolutions of two wheels working together are governed entirely by their diameters. A wheel one-half the diameter of another making twice the revolutions. The number of teeth in these wheels must also be in exact proportion to their diameters.

The following rule should be carefully studied, viz.: "The primitive diameter of a wheel is to the primitive diameter of a pinion as the number of teeth in the wheel is to the number of leaves in the pinion." For example:

We have a wheel of 80 teeth and a pinion of 10 leaves. The primitive diameter of the wheel is 16 m-m required the primitive diameter of the pinion.

$$80 : 10 :: 16 : (x)$$
$$x = 2 \text{ m-m}$$

45

In finding the sizes of lost wheels or pinions, we have no sizes to start with; in many instances we do not even have the number of teeth in the required wheel; we first find the correct number of teeth by counting up the train, which will be more fully explained later, then by carefully measuring our distance between centers by the aid of a depthing tool, we can easily and accurately find the correct sizes by the following rules:

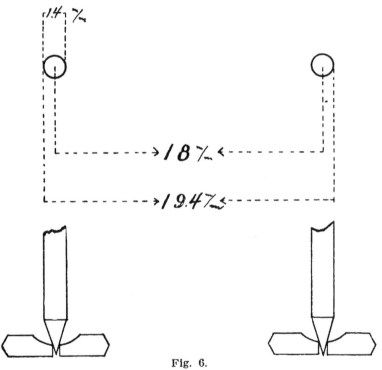

Fig. 6.

The distance between centers is the distance from the center of a wheel to the center of the pinion into which it depths, but we must take our measurements from the plates from jewel to jewel. The points of the depthing tool should be set so they enter the centers of the two jewels while the spindles of the tool are perpendicular to the plate, as shown in Figure 6. To find the exact distance from center to center, we measure from outside of one spindle to the outside of the other, and then deduct the diameter of one spindle (both being the same), the result will be the exact distance between centers. This is also clearly shown in Figure 6. Comparatively few of the best watchmakers seem to be familiar with the millimeter gauge. It is so much ahead of the old degree gauge and less cumbersome than the micrometer, that I would advise all of my readers to get one, and become thoroughly acquainted with its use; when this is done, am sure you will wonder how

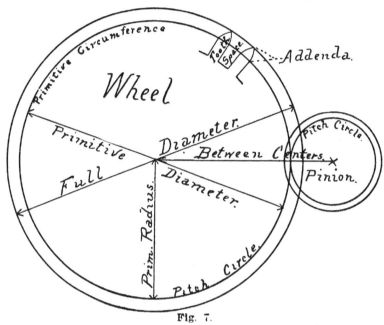

Fig. 7.

you ever were able to get along without it. All measurements referred to in these articles will be from the metric system.

The gauge referred to will measure 1-250 of an inch, and a person familiar with its use can measure as accurately to 1-500 part.

The terms that are shown in Figure 7 should be well understood, and are explained in the following:

1. The *full diameter* of a wheel is the distance from the point of a tooth on one side to the point of a tooth on the opposite side.

2. The *primitive diameter* is the full diameter less the addenda. It is also the distance from the primitive circum-

47

ference (or pitch circle), through the center to the primitive circumference on the opposite side.

3. The *primitive radius* is one-half of the primitive diameter.

4. The *addenda* is the difference between the full and primitive diameters.

5. The *distance between centers* is from the center of the

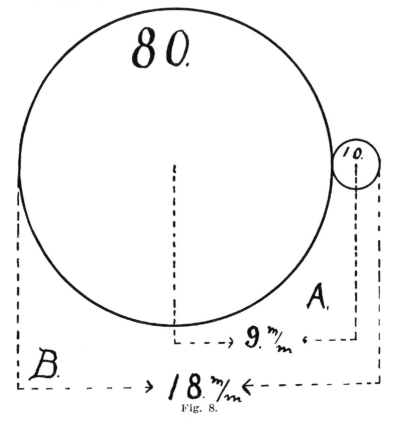

Fig. 8.

wheel to the center of the pinion, and is *always* one-half of the sum of their primitive diameters.

6. The *circular pitch* is found by dividing the primitive circumference by the number of teeth or leaves.

7. The *diametrical pitch* is found by dividing the primitive diameter by the number of teeth or leaves

WHEELS AND PINIONS.

8. The diametrical pitch × the number of teeth or leaves = the *primitive* diameter.

9. The primitive diameter + the addenda = *full diameter.*

10. The sum of the primitive diameters equals twice the distance between centers.

11. Twice the distance between centers ÷ the sum of teeth and leaves of the pair working together gives the *diametrical pitch.*

12. The *full diameter* of a train wheel is the *primitive diameter* + *two and a half* (2½) diametrical pitches.

13. The *full diameter* of a pinion is the primitive diameter + one and one-fourth (1¼) diametrical pitches.

In Figure 8 the distance between centers (b. c.) is shown by A or 9 m-m, twice the distance between centers or the sum of their primitive diameters is shown by B or 18 m-m.

We will proceed as follows to find the full diameters of both wheel and pinion.

9 m-m × 2 = 18 m-m twice b. c.

80 + 10 = 90 sum of teeth and leaves.

18 (sum of diameters) ÷ 90 (sum of teeth) = .2 m-m diametrical pitch.

.2 (d. pitch) × 80 (no. of teeth) = 16 m-m primitive diameter of wheel.

.2 × 2½ = .5 addenda for wheel.

16 m-m + .5 m-m = 16.5 m-m full diameter of wheel.

.2 × 10 = 2. m-m prim. di. of pinion.

.2 × 1¼ = .25 m-m addenda of pinion.

2 m-m + .25 m-m = 2.25 m-m full diameter of the pinion.

We find the sizes of all wheels or pinions in the same manner only allow different amounts for the addenda, for instance in the stem wind wheels, we add *two* diametrical pitches to the primitive diameter.

For train wheels, we will add two and a half for the wheel and one and one-fourth for the pinion.

Our work can be greatly simplified if we understand the principles involved by working out our problem as follows:— the diametrical pitch will be found as before.

To find the full diameter of the wheel, multiply the diametrical pitch by the number of teeth + 2½.

80 + 2½ = 82½.

.2 (diametrical p.) × 82½ = 16.5 m-m full diameter of wheel.

To find the pinion multiply the diametrical pitch by the leaves + 1¼.

10 + 1¼ = 11¼.

.2 (diametrical p.) × 11¼ = 2.25 m-m full diameter of pinion.

In actual practice, we obtain the same results with considerable less work.

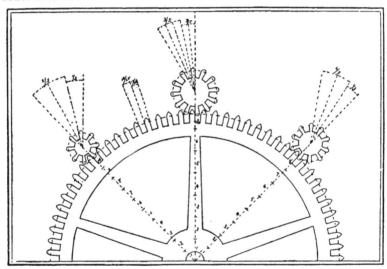

Fig. 9.

I will give another problem to illustrate the method more clearly. Suppose we are repairing a watch, and have reason for believing the depth between the third wheel and fourth pinion is faulty, we can tell very quickly which is of improper size; with our depthing tool we find the center distance to be 8.5 m-m. The wheel has 75 teeth and the pinion 10 leaves.

We first double the center distance.

8.5 m-m × 2 = 17.0 (Twice b. c.).

75 + 10 = 85 (sum of teeth and leaves).

17 ÷ 85 = .2 diametrical pitch.

WHEELS AND PINIONS.

Wheel $= .2 \times (75 + 2\frac{1}{2}) = 15.5$ full diameter.

Pinion $= .2 \times (10 + 1\frac{1}{4}) = 2.25$ full diameter.

These sizes will be the correct ones, and by comparing the sizes of the wheel and pinion with the ones just found any defects can be detected.

I would advise my readers who are interested in this very important part of their work to take down some movement, and figure out the sizes of several pairs of wheels and pinions, as by so doing the principles will be better understood and will be less liable to be forgotten.

The primitive diameters of all wheels and pinions working together are in the exact proportions to each other that their teeth and leaves are. This will be better understood from Figure 9. Here we have a wheel of 80 teeth, only half of it being shown, three pinions depth into this wheel each having a different number of leaves; they must therefore have different diameters; the one at the left has eight leaves, which is one tenth ($8 \div 80 = 1\text{-}10$) of the number of teeth in the wheel. The primitive diameter or radius of this pinion must be exactly one-tenth that of the wheel, which is clearly shown by the divisions showing the radius of the wheel divided into ten parts.

The pinion on the right has ten leaves; its radius then must be one-eighth that of the wheel as shown by the eight parts, while the size of the pinion in the center having twelve leaves would be found by dividing the diameter of the wheel into $6\frac{2}{3}$ parts. $80 \div 12 = 6\frac{2}{3}$.

The eight and ten leaf pinions have a ieaf and space divided into three parts, the leaf having one part, and the space two parts; while with the twelve leaf pinion, a leaf and space is divided into five parts, two for the leaf and three for space.

In the wheel the spaces are a trifle wider than the teeth, 13-25 being allowed for the former and 12-25 for the latter. The addenda is not always the same. We allow two and a half diametrical pitches for train wheels, while for stem wind wheels we allow only two. This makes the teeth a trifle shorter.

The construction of the train determines the vibrations of the balance in a given time, or as we commonly speak of it,

51

18,000, 16,200 or 14,000 vibrations per hour, the first one being the fast train now in common use, and the last two the slow trains which are not used to any great extent except in marine chronometers.

Nearly all watches made now have the fast train with 300 vibrations a minute. The escape wheels have fifteen teeth (the rule is so general that we will consider it an exception

Fig. 10.

when any other number is used), and as each tooth of this wheel gives impulse alternately to the R. and L. pallet, it must give twice the number of impulses or vibrations to a revolution as there are teeth in the wheel (15 × 2 = 30). If there are 30 vibrations of the balance to one revolution of the escape wheel, then there must be as many revolutions of the escape wheel per minute as 30 is contained in the vibrations per minute:

300 ÷ 30 = 10) fast train.

270 ÷ 30 = 9 ⎫
⎬ slow trains.
240 ÷ 30 = 8 ⎭

WHEELS AND PINIONS.

Should we wish to replace a hair spring, we can quickly determine the number of vibrations the watch should have by dividing the number of teeth in the fourth wheel by the number of leaves in the escape pinion, and multiplying this result by twice the number of teeth in the escape wheel, for example:

The fourth wheel has 80 teeth.

The escape wheel has 15 teeth.

The escape pinion has 8 leaves.

$80 \div 8 \times (2 \times 15) = 300$ vibrations per minute.

Fig. 11.

In all cases where there is a second hand the above rule may be used. In case there is no second hand, then we must start from some point that makes a definite number of revolutions in a given time, as the center wheel, which makes one revolution each hour. In the article on Hair Springs and Springing, this will be more fully explained.

The train of a watch is so sensitive, and the motive power so very small, that it requires but a slight imperfection in the construction, or a very small amount of wear to cause it to stop. We often find a train that to all appearances is perfectly free, yet the watch will stop, and before we have a chance even to examine it, it will start off and perhaps run for hours before it will stop again. We find this often in movements that have been running for a long time and are badly worn. The trouble is frequently found in the escape pinion for two reasons, first it revolves more rapidly than any of the others and is more liable to wear, and, second, it requires less power at that point to stop the train than at any other. Figure 10 is a photograph of an escape pinion highly

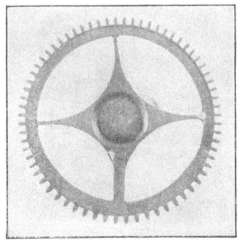

Fig. 12.

magnified, showing where the fourth wheel has worn a leaf half in two. This has the same effect as a shallow depth, and the watch often stopped until a new pinion was put in, after which the trouble disappeared. In some cases we can raise or lower the fourth wheel so it will work above or below the worn place, which will overcome the difficulty.

Some of our best lessons may be learned by the photographs which will be shown from time to time. These will show how work should be done, and also how it should not be done. We repair a watch, but as we place it to our ear we hear it "grind," examine the pivots and they appear in good condition, but place the same pivot under the compound microscope and examine it thoroughly, and I do not hesitate a moment in saying we will all try and do a little better next time. A well polished pivot looks slightly rough, but when we examine those made by a careless workman, it is no wonder so many watches fail to perform their duty in a satisfactory manner.

Often a watch is improperly oiled, and the pivots get dry

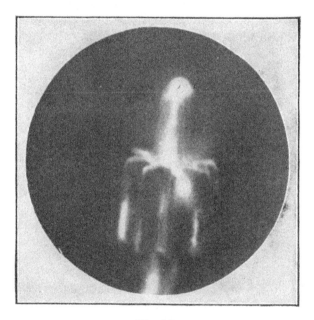

Fig. 13.

and begin to cut in a short time; they are so badly worn that a new pivot is necessary. Figure 11 shows such a pivot, where it is worn nearly half way off. When the surface becomes dry, a powder soon forms, which seems to cut the steel very rapidly, and soon the pivot will bind in the plate, if not jeweled, and in some cases they can not be removed without breaking.

Many very good workmen do not think it is necessary to understand how to find the correct sizes of wheels and pinions as shown in these articles, but I am sure that one who does understand how to do it intelligently will have much the best success with his work. How often we see a workman "round up" a wheel where the depth appears faulty which is caused by the wheel being staked on the pinion out of center; the wheel may be the exact size for the pinion, yet being out of true in the round, caused by being decentered, the depth would be too deep on one side and too shallow on the opposite side; by "rounding up" the wheel its size will be reduced and the wheel will be too small for the pinion.

The proper way would have been to cement the wheel on a chuck in the lathe and true from the outside of the teeth with peg wood, cut out the center true and bush to fit the pinion. In this manner the wheel would retain its correct size, and yet be perfectly true.

To illustrate the necessity of understanding this branch of our work thoroughly, I show in Figure 12 a photograph of a wheel taken out of a French clock which would not keep time. It had been repaired, but failed to work properly. A portion of this wheel had been broken and a section of another wheel was soldered on the original wheel. This might not have been so bad, but the wheel in the first place had 72 teeth, but the portion replaced had coarser teeth, and the wheel as repaired contained 68 teeth instead of 72, which it should have had; this being the center wheel and making one revolution each hour, the clock, if it run, would gain constantly. The original wheel having 72 teeth and the repaired one 68, it would gain ($72 \div 68 = 1$ hour, 3 min., 31 13-17 sec.). Each hour the clock would gain 3 minutes and 31 13-17 seconds, or about 84 minutes a day. Figure 13 is another very peculiar case that recently came to my notice. The pivot of the escape pinion in a finely adjusted watch was broken, the one who repaired it not being able to properly replace a new pivot, filed the arbor of the pinion to a point and tried to make it run in the jewel. It was not a success, and a new pinion had to be turned to replace this one, after which the watch kept good time again.

I often wonder how any one who has any conscience can do such work, and trust that these examples may be an incentive for all to do better work.

THE BALANCE STAFF AND ITS MEASUREMENTS.

There is no part of a watch that requires greater care in its making than the balance staff. Much of the close timing depends upon the accuracy and fine finish of the pivots of this delicate part of our watch.

Doubtless many of my readers will say, it is not necessary to know how to turn a staff; that it is time thrown away in learning to do it, as we can buy them very nicely finished which have been made by automatic machinery, and perhaps are better than the ordinary watchmaker can ever hope to make them with the small amount of practice he has in that line to-day. They can be bought for less money than the time is worth that must be spent in making them, yet I must insist that it is absolutely necessary for the watchmaker of to-day to know how as well as those of a quarter of a century ago, before the introduction of automatic machinery.

We may be able to buy a staff that will fit the watch we have to repair nine times out of ten, but the tenth time we are unable to do so, and we will be obliged to make it ourselves. It is just as important to *know how* to make one well if we only have it to do occasionally as it would be if we were making them constantly, possibly more so.

We are constantly getting watches to repair that no ready made pieces will fit, viz., those made in foreign countries. The Americans were the first to make their movements with interchangeable parts, so duplicate parts could be obtained and quickly fitted. The foreign manufacturers were very slow to see the advantages of this system, and, as many of their watches are made at the homes of the watchmakers instead of in factories, it is easily seen that their parts could not be interchangeable, and getting duplicate parts to replace broken or worn out ones becomes quite a serious matter to the inexperienced workman. Many of the very finest Swiss watches have no duplicate parts, therefore each individual piece must be made for that particular movement. For this

57

reason the successful watchmaker of to-day requires as great skill as those of a quarter of a century ago, although he may not be required to show his skill so often in some ways as

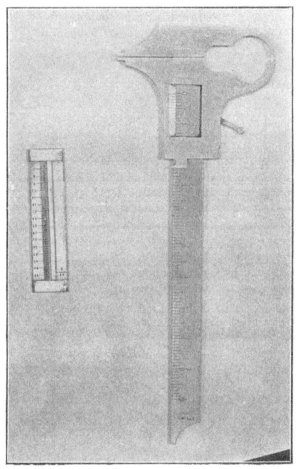

Fig. 1.

formerly. Unless we are able to make our repairs as good as the original, our work is never properly done.

To make a balance staff properly, requires good judgment, patience and skill. We should never depend upon the old staff for our measurements, as in many cases the old one

THE BALANCE STAFF AND ITS MEASUREMENTS.

is wrong, and we would make a new one with the same faults. We should take them from the watch itself, making each part of the staff to correspond. This may seem a very hard thing to do for those who have not been in the habit of doing so, but it is a very simple thing to take them accurately and quickly by the use of the millimeter gauge shown in Fig. 1. I would advise every workman to add this useful tool to his equipment, if he has not one. Its use will be easily mastered.

We often find a balance cock that is badly bent, or one that has several burrs thrown up with a graver to *adjust end shake*, etc., etc. Our first step should be to restore all parts to their original position or condition before attempting to

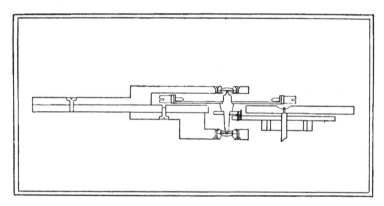

Fig. 2.

make our measurements. When this has been done, we will first measure from the *outside* of the top cap jewel to the *outside* of the bottom cap jewel. From this measurement take the thickness of both cap jewels, and we have the distance from the inside of one cap jewel to the inside of the other, or the exact length of the staff, making no allowance for end shake, which should be just enough to be seen, about one-half of one-tenth of a millimeter.

We obtain the height of the roller by getting the distance from the outside of the bottom *hole* jewel to the top of the lever, adding to this enough for clearance, about 2-10 m-m *and the thickness of the roller.* The balance seat is located in various ways, in a full plate watch like the one shown in Fig. 2. We may get the distance from the outside of the bottom hole jewel to the top of the upper plate, adding just enough for clearance, about 2-10 m-m.

In cases where the hair-spring stud is stationary and cannot be raised or lowered, a difficulty presents itself; if we get our balance a trifle too high, our hair-spring will be high in the center, and should it be too low, the spring will be low in the center. All watches of this kind, I get the distance from the outside of the top hole jewel to the underside of the hair-spring collet, while the stud is in place on the balance cock and the spring level with it; to this measurement add

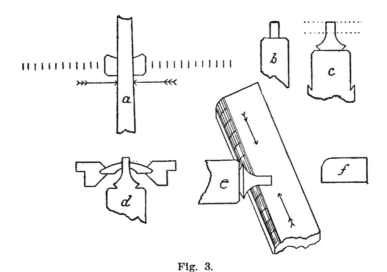

Fig. 3.

the thickness of the balance arm and we have the exact distance from the end of the top pivot to the shoulder for the balance.

The hub of the staff should be slightly less in diameter than the width of the arm of the balance and have a graceful taper; the roller should fit squarely against the hub of the staff, and should require but a slight tap of the hammer in staking it on. It is a mistake fitting them as tightly as is often done; by so doing the staff is liable to become bent and unfit for use.

The shoulder beneath the roller may be undercut a trifle, but it should not be done in a manner that will weaken the staff.

In fitting the hair-spring collet, it is a bad practice to try on the collet itself, but a better way is to take the measurement by slipping the collet on a smooth broach as shown

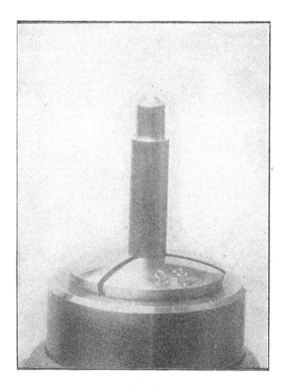

Fig. 4.

(A. Fig. 3), and getting the size of the broach at the points shown by the arrows. The taper of the broach will be just enough to make the collet fit nicely.

Nearly every workman thinks his way is the only proper one, perhaps because he is more familiar with it. My advise is to learn all the ways possible, and then use the best points of them all in your own work, being ready at all times to drop an old idea for a modern or better one. We must be *progressive.*

The first step toward making a staff after our measurements have been taken, is to prepare our steel. We can buy

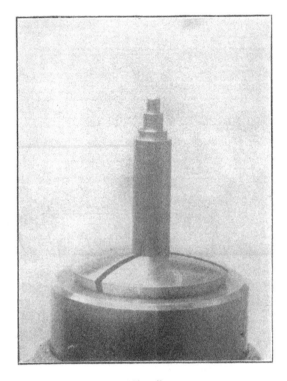

Fig. 5.

blanks hardened and tempered ready for use, but I am in favor of preparing my own steel, as much depends upon this important part, as has already been demonstrated in the previous article on Iron and Steel. That which has been overheated in hardening will be very brittle and unfit for our work; only that hardened at a very low temperature, barely a cherry red, will make the best staff wire. After hardening, we can polish the surface and draw the temper to a purple blue, or a better way, is to take an old iron spoon, place the hardened wire in it with a small piece of

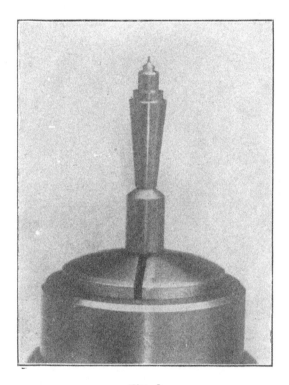

Fig. 6.

beeswax or a little lard oil and heat over a lamp until the beeswax or oil ignites. This will make a hard staff; if you allow it to burn off, it will make a softer staff, one that will turn easily. It is a mistake, making a staff too hard, as it wears no better, and is more liable to break.

Some workmen prefer to turn the lower end and some the top end of a staff first. In my own work, I always turn the top end first. This may not be the best, but to me it seems that way.

The wire should project from the chuck of the lathe far enough to make the whole staff before we begin our work;

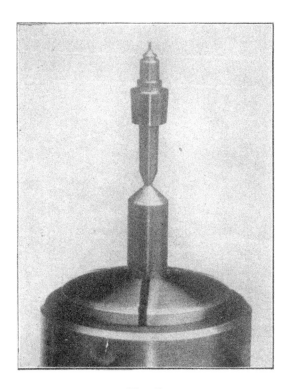

Fig. 7.

our first step is shown in Fig. 4, where the shoulder has been turned to fit the balance arm, and also the correct distance from the end. Fig. 5 shows the second step—the hair-spring collet has been fitted, the undercut made for riveting over the balance arm and the part turned down for the pivot. This should never be over two-thirds the size of the collet shoulder. Fig. 6 shows the pivot and the back cut complete with the hub blocked out ready to grind and polish. The whole staff is now nicely ground and polished except the balance shoulder. First, we grind with oilstone powder and oil, being careful to always grind in a diagonal direction in order that we may get a perfectly smooth surface. The staff is now thoroughly cleaned and ground again with crocus and oil in the same manner, and after cleaning again is polished with diamantine and oil, using just enough oil to make a very

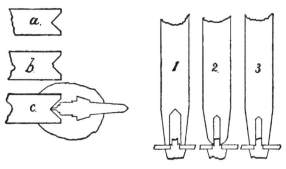

Fig. 8.

thick paste, as it will not polish until the oil becomes quite dry. The object of turning the staff to a long taper as shown in Fig. 6 is to allow us to grind and polish the hub perfectly flat, which could not be done if we should block out the hub first, then the corners would be rounded, which ruins the beauty of fine work. After the polish is completed, we may block out our lower end as shown in Fig. 7, which shows it ready to be broken off and placed in the cement chuck, as shown in Fig. 8 at c; at a is shown a cement brass properly centered; at b is shown one as they are liable to be. It requires considerable practice to be able to center a piece of brass perfectly, but this is necessary in order that our finished staff should be perfectly true.

Our cement brass should be heated hot enough to melt the cement in order that it may hold properly; the staff should also be hot, and while warm should be trued up with a piece of peg wood. The lathe should be kept in motion until the cement gets set, or it would not remain true, if the cement should be thicker on one side than on the other. The staff will often be thrown out of true by the unequal contraction in cooling. In such case, turn the cement round, reheat and true, when it will remain so.

Cone pivots must be perfectly cylindrical where they enter the jewel; in fact, they are the same shape on the end as a square shoulder pivot, the cone shape being used to give greater strength. This is shown in Fig. 3 at *b* and *c*; at *d* is shown a pivot as it should be fitted to a jewel. The pivot should *always* be long enough to go more than through the jewel, so the cap jewel will force it back slightly; in this way there is no danger of the cone ever binding in the jewel hole. This causes many watches to lose motion in different positions. The end of the pivots should be ground perfectly flat and nicely polished. It is a good practice to polish the end before turning down the pivot; this insures its being perfectly flat.

The pivot and the cone must be ground diagonally, as shown in Fig. 3 at e, the grinder and polishers being moved in the direction of the arrows; a cross section of the grinder and polishers is shown at *f*, the corner being rounded to fit the cone of the pivot. Fig. 9 is a photograph of the iron grinder and the bell metal and tin polishers. The shape of all is the same, the one at the right showing the rounded corner used for cone pivots; near the end where the rounding is the greatest, it can be used for large pivots, and farther back for small ones, there being some place that will fit any pivot. We use the oilstone powder on the iron, the crocus on the bell metal and the diamantine on the tin polisher.

Fig. 8 shows the method of staking on the balance wheel; punch 1 is a hollow flat one, large enough to fit over the staff and drive the arm down to the shoulder; 2 is a hollow round face, which fits closely and spreads the undercut; 3 is a flat face that rivets it down perfectly smooth.

Fig 1 shows, besides the millimeter gauge, a pivot gauge, which is a great help in fitting a pivot correctly. By having a set of pivots turned on the end of pieces of wire, we can select the one that fits our jewel properly, then determine

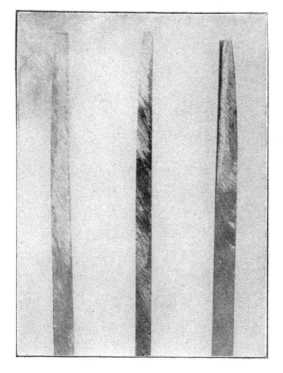

Fig. 9.

its size on the pivot gauge, and turn our pivot to that size, allowing the gauge to hang lightly on the pivot.

In making the back cuts or under cuts, it is necessary to have the gravers perfectly sharp and touch the work very lightly, as a trifle too much pressure will break off the delicate point, and prevent the gravers from cutting.

Some object to the use of cement, as it is so hard to clean off. This is easily overcome by placing our staff, after removing from the cement, upon a small piece of brass or copper, and heating slowly over a lamp. Most of the cement will melt off, and only a very little alcohol will be needed to dissolve that remaining.

A splendid lathe cement can be made by melting ordinary shellac, and adding a small amount of balsam of fir. This makes a cement that is much stronger than ordinary cement.

JEWELING.

There is no part of the watchmaker's art that requires more skill, in its performance, than that of jeweling, and yet from the many specimens of poor work that are seen so often, one would think that knowledge upon this subject was very meager. We see jewels cemented in with shellac, others that are loose in their settings, and in many cases jewels that are set in upside down, and at times, those set are so much out of flat that the pivots will pass diagonally through them.

Jewels are used for two purposes, to reduce the friction and make a harder surface for the pivot to act against, thus lessening the wear and consequent side shake of the pivots, also requiring less motive power to run the watch, and as we reduce the strength of the main spring, we also reduce the wear throughout the train, giving a longer life to the movement.

Jewels are made of various substances, ranking in quality according to their hardness, the sapphire and ruby being the best, and the more common ones are made of garnet, while the cheapest ones are nothing but glass, and are worse than none. The diamond is sometimes used for the cap jewel in some of the better grades of movements, and on account of their extreme hardness and their high polish, are very valuable for that purpose.

Jewels are very helpful, when properly set, but are worse than none when poorly set, or when those that are used are not well polished; a good, brass bearing for a pivot, when nicely burnished with a smooth broach, makes a splendid substitute for a jewel, and will wear for many years. We see many of the finest chronometers, costing several hundred dollars, where the greater portion of the train is not jeweled at all.

Figure 1 shows a cross section of the jewels in common use; A and B are two forms of cap jewels or end stones; C, an American plate jewel; D, a Swiss plate jewel, also used as a balance jewel at times in the cheapest work; E, a balance jewel used in the best work.

Figure 2 shows a cap jewel and a balance jewel in their settings with a pivot passing through the hole jewel, resting

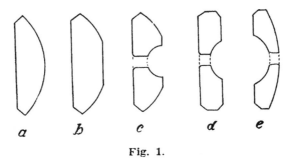

Fig. 1.

against the cap jewel. The thickness of the balance jewel should be such that the friction of the pivot on the side and that on the end would be the same when the watch is in different positions.

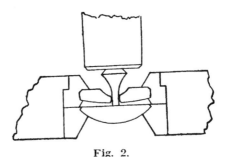

Fig. 2.

Jeweling is not a difficult task, and but few tools are required to do it nicely; two or three jewel-gravers (Fig. 3), one or two jewel burnishers (Fig. 4), and a burnish file are about the only tools necessary, it being expected that the

70

average watchmaker is well supplied with a good assortment
of drills, brass wire, etc.

Jewel gravers should be made triangular in form with the
lower corner removed as shown in the end view of Figure 3.

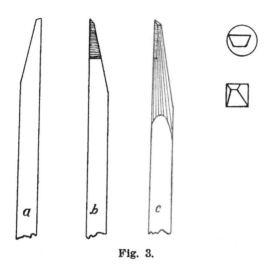

Fig. 3.

The top of same graver is shown at A, the side at B, and the
bottom at C; these can be easily made from wire about 4 m-m
square, hardened and tempered to a dark straw.

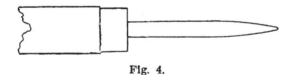

Fig. 4.

I will explain first how to set the jewel and make a new
setting to replace a broken plate jewel; often new jewels are
set in the old settings. This is far from satisfactory, although it can be nicely done if the new jewel is slightly

71

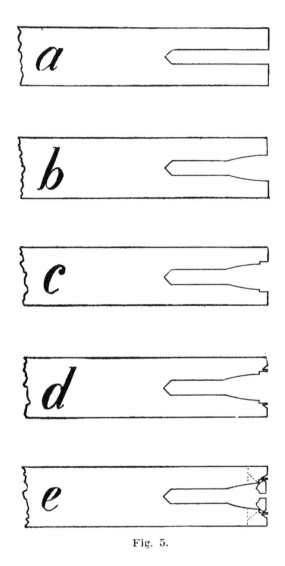

Fig. 5.

larger than the old one, as we can cement up the old setting, cut a new seat and fit the jewel properly; where the jewels are set in gold, it is often desirable to use the old settings again. The following remarks apply to either new or old settings, only we begin by setting the jewel in the wire in the former case.

The first step will be to center the wire and drill the hole A (Fig. 5). This hole should be about two-thirds as large as the diameter of the jewel to be set. We can never depend upon a drill making a true hole, so after drilling we cut out the hole perfectly true with the jewel graver as shown at B until about three-fourths the diameter of the jewel. We are now ready to cut the seat for the jewel as shown at C, making it just large enough that the jewel will fit loosely in the hole, yet without perceptible side shake. Many good workmen try to cut the seat for the jewel to rest against the same curvature as the jewel by using a jewel graver with a rounded end; theoretically this is correct, but in actual practice it is hard to do. I think it is better to make a square shoulder, as shown in cut C (Fig. 5), allowing only a very small shoulder for the jewel to rest against, then stripping out the setting just to this shoulder as shown by the dotted lines in E (Fig. 5).

Before burnishing in the jewel, it should be just below the surface of the wire. This can be done by burnishing the wire at the edge of the hole over the jewel or by cutting a bezel around the hole and burnishing the thin brass over the jewel as at D and E (Fig. 5).

The burnisher (Fig. 4) should be very hard and highly polished; in making one, it is best to leave it full hard, not tempering at all, otherwise the brass will adhere to the burnisher, and prevent nice work; a little beeswax will often help the burnisher to perform its work in a more satisfactory manner. When the jewel is perfectly solid, the end of the wire may be faced off until nearly level with the jewel. It should now be tested to see if it is true. We find many jewels where the hole is not in the center; when such a jewel is set, the hole will not run true as the outside of the jewel fits the hole in the wire, the result is the hole will not run true in the lathe. Should we now turn down the outside to fit the plate,

73

the hole would not be in the center of the setting. We test the trueness of the jewel with a long piece of pegwood, the point of it being in the hole of the jewel, and allowing it to rest on the T rest near the point. This will magnify the motion at the other end of the pegwood, and we can quickly determine its trueness. Should it not be true, we should cut off the wire at once, long enough to make the setting, after which we will cement up the jewel, truing from the

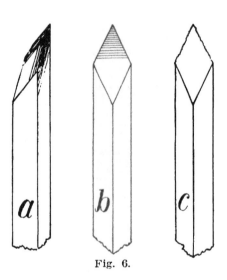

Fig. 6.

hole by means of a piece of pegwood. Now, by turning down the outside of the setting, it is easily seen that the hole of the jewel must be in the center of the setting. After the setting is turned to fit the plate, it can be reversed and the stripping done. This is a very difficult thing for most workmen to do. There is a little "knack" in doing it, that must be learned by experience, but the most important thing is to have a graver in proper condition. It must have a long point, sharpened perfectly flat, and be nicely polished. The polishing should be done by drawing the graver on the polisher parallel to the cutting edge, as shown in Fig. 6, at A; this leaves the edge of the graver very smooth, enabling us to cut a polished surface while a graver polished crosswise, like B, leaves the edge very rough, like C, making it impossible to make a smooth cut. This is very important; about the best polisher for our gravers is a piece of the plate glass about four or five inches square with a piece of 4-0 or 5-0 emery paper cemented upon one side of it, the glass being hard allows us to polish the gravers without rounding the edge. In using the graver for the final cut, we should either give it a slightly drawing or sliding cut, otherwise the surface is liable to be full of rings, and not nicely polished. The setting should not be polished with rouge or any other polishing material, as that has a tendency to round the corners, which ruins nice work. The top or flat surface of a setting seems to be the most difficult part to finish for the ordinary workman, yet it is not difficult. We require a piece of finely ground plate glass; a piece about three or four inches square is a convenient size; in fact, it is best to have two of them. The plate glass can be obtained from any dealer in glass for a few cents, and may be ground with emery flour and water, grinding them with a circular motion until the whole surface is well ground. Another necessary tool is the burnish file, which is never in proper condition when bought, but must be carefully prepared for use. One side is ground flat and quite smooth, and should be refinished by drawing the file crosswise on a piece of rather coarse emery paper or a No. 1 emery buff stick. This leaves the surface of the file in very fine lines from side to side. After the surface of the file has been prepared, we must be very careful not to touch it with our fingers, as absolute cleanliness is necessary for nice burnishing, a finger mark or a particle of oil being enough to prevent the surface of the setting from polishing.

As in many other things, there is a little "knack" in jewel burnishing that can only be acquired by experience. We proceed as follows: The setting is first ground upon the ground glass with a little tripoli and oil or oil stone powder and oil until it is perfectly flat and is just thick enough to be flush with the surface of the plate. It is now thoroughly cleaned with a little benzine, which removes the oil, then with a piece of clean pith. It is now ready for the final burnishing, the file having been previously prepared and perfectly clean. We place the jewel setting on it with a clean pair of tweezers (the fingers should not touch the jewel or

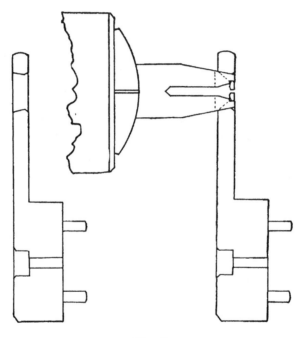

Fig. 7.

file at all during the progress of the work). Carefully place a piece of tissue paper over the setting, place the fingers upon it, bearing very heavily upon the paper and setting, give a few strokes lengthwise of the file with considerable pressure and the job is done, and should be as well polished as those in a new movement. Unless our repairing is finished as

76

nicely as the original pieces, we are not doing our work as well as it should be done.

In setting either plate or balance jewels, we should set the cup side in, having the flat side out if a plate jewel, and the convex or rounded side out if it be a balance jewel. The convex surface of the balance jewel should be slightly below the surface of the setting, so the cap jewel, or end stone, will not quite come in contact with it, as clearly shown in Fig. 2. There should always be a very small space between the balance jewel and the cap jewel. Formerly the balance jewels had a flat surface like the cap jewels, but the convex surface is a great improvement on account of the oil being held between them by capillary attraction, and is gradually fed to the pivot, as the jewels are close together at that point. It is a very bad practice to have the cap jewel loose and resting directly upon the balance jewel, as we so often see them in Swiss watches.

In cleaning watches, we should carefully examine the end stones, as they often become pitted very badly, greatly interfering with the time keeping qualities of the watch. Many of the best end stones as now made are flat on both sides, the top in many cases being flat just in the center. When we find a jewel of this kind that is pitted, we can take it out of the setting and reverse it in the same setting. This presents a new surface to the pivot, and is as good as a new jewel. In many new watches the jewels are set in this manner, the end of the pivot acting upon the small flat surface instead of the large surface as is commonly the case. It is clearly seen that by setting the flat surface against a square seat in the wire, we can get the jewel perfectly flat, quite an important point with an end stone.

The pivot that has worn a pit into the cap jewel should be ground and repolished upon the end in every case where the cap jewel has become pitted. It is not generally known that the jewel which is being pitted charges the end of the pivot with its powder, the softer of two materials will always become charged, the principle being the same as a piece of steel, which is charged with diamond dust, being used to grind a diamond. I have seen several cases where new cap jewels had been put in to replace pitted ones that in a very short time were as bad as the ones that they replaced, simply because the pivots were charged with the powder, and had not been refinished. I have seen several cases where the pallet stones were very badly worn where the teeth of a brass escape wheel came in contact with them, while the teeth of the wheel were not worn enough to be noticed. These teeth had become charged the same as the pivots, and it was necessary not only to replace the pallet jewels, but also the escape wheel. The principle is much the same as it would be to try to grind a piece of lead on an iron or steel lap with emery powder. The emery would imbed in the lead, and this would form a lap that would cut the iron or steel very rapidly, and not the lead.

Perhaps our most difficult piece of jeweling is to set one in a Swiss bridge after the bezel has been ruined, and there is no way of burnishing in the jewel. This can be done by bushing the bridge and setting the jewel in the bushing. First, we should place the lower plate of the movement in a face plate and true from the jewel hole in it. When this is done, the bridge to be jeweled is put in place and screwed on solid, having trued from the lower hole; the top one must also be true, as this is our method of uprighting. The hole will be cut out slightly tapering, as shown in Fig. 7, just to the edge of the previous stripping, slightly countersinking the underside as shown. The next step will be to remove the bridge from the plate, turn down a brass wire so its taper will perfectly fit the taper of the bridge, allowing the end to come through just enough so we can burnish it over the bridge where it has been countersunk. The brass will now be perfectly solid, and the bridge will revolve with the wire in the lathe. The hole in the bridge was first cut out perfectly true, then the wire was turned true, and the bridge burnished securely upon it, both must now be true, as they each have the same center. By centering the wire in the bridge and drilling the hole, we can set the jewel in the same manner as in the end of the brass wire, it only being necessary to set it the right depth to make the end shake correct, which may be easily determined with the millimeter gauge. By a careful study of Fig. 7 the method of finishing the upper side of the bushing will be understood. First, the wire is cut off at the dotted line just above the finished part of the bridge, then screwed in place on the lower plate, which has not been removed from the face plate. If our work has been well done the hole in the jewel we have set will revolve perfectly true in the face plate, and we can strip out the bushing to the dotted line, which cuts off the projecting part of the bushing, leaving our work as good as new. I have seen bushings put in brass plates that were nickel-plated, the bushings being made of nickel wire, and so nicely done that none of

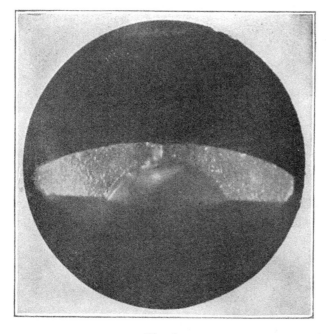

Fig. 8.

the brass showed at all; in fact, the most careful examination would not show that a bushing had been put in. It should be our aim in all work done to make it so perfect that repairs of any kind cannot be detected.

There are many other kinds of jewels, such as the pallet stones, jewel pins, etc., which are not spoken of at this time, but they will be explained fully in other articles on the lever escapement.

Figure 8 is a balance jewel ground through the center showing the bearing of the pivot. It is highly magnified, and is a very good representation, the photograph having been taken from a very fine sapphire jewel. The balance jewels have "olive holes," as shown in this photograph, and also in Fig. 1 at E. The friction will be less in such a hole than it would be in one where the hole was the same size all the way through.

PIVOTING.

About the most difficult task a watchmaker is required to do is that of pivoting. Pivots are broken, or rendered unfit for use, in many ways. When a watch falls, or strikes any hard substance with considerable force, a pivot is liable to be either broken or badly bent; in either case the watch will be unable to perform its duties as a timekeeper. If the pivots are not properly oiled and become dry they soon wear so badly that the only remedy is a new pivot; then again, we find those that have been pivoted, but so poorly done that the watch cannot perform well.

When one observes how some workmen resort to almost every conceivable method to avoid the task of pivoting, the natural conclusion would be that it is a very difficult task and but few are able to perform it in a satisfactory manner.

To say it is an easy task to pivot properly, especially today, when the staffs, pinions, and other steel parts are left so hard, would be far from a truthful statement, as it is one of the most difficult tasks that comes to the average watchmaker. Let me repeat here what has been said before, unless our work when completed is as good as the original and does not show where it was done, it is not properly performed, and should not be used in good work.

How often we see pivots not in the center of the staff, or pinion pivoted, caused either by the carelessness of the workman, or by the chucks of the lathe being out of true. Again, some cases where the work has been fairly well done, but previous to doing it the temper had been drawn until the body of the staff, or pinion, and sometimes, even the arms of the balance wheel, were drawn to a blue. All staffs and pinions cannot be drilled without drawing the temper, but when this is necessary all traces of the blue should be removed, either by polishing with diamantine and oil, or by some good blue remover; one that acts well and quickly is made as follows:

Aromatic sulphuric acid,
Sweet spirits nitre, } Equal parts.

This when applied to the steel parts that have been blued will remove the color very quickly, but in all cases where it is used great care must be used to remove all traces of the acid, otherwise the steel is liable to rust. By washing the article in dilute ammonia, after which it is placed in grain alcohol a moment, and, lastly, dried in box-wood sawdust, no danger of rust need be feared.

We should never draw the temper until we find it impossible to drill without, although most American staffs and pinions are too hard to be drilled without drawing, as it is frequently necessary to do so we should adopt some method that will do the work quickly and easily with as little harm to the other parts as possible. There are a great many appliances on the market for this purpose, but few of them seem to meet our requirements. Copper should be used, as it is one of the best conductors of heat we have. The form is also quite important. Some use a piece of copper bent in the form at a, Fig. 1. This does the work fairly well, but has the disadvantage of not completely surrounding the staff, the heat being directly conducted only on two sides, and the ends being square are liable to draw the arms of balance as well. About the best method is to have several pieces of copper wire with various sizes of holes drilled in them which will closely fit the part of the staff, or pinion, to be drawn. By sawing into the end through the center of the hole the wires may be closed enough to clamp the work firmly. The end of the copper wires should be rounded as shown at b, Fig. 1. At c is shown the wire in place on a staff. It will be noticed that by rounding the end of the wire it does not come in contact with the arm of the balance and is less liable to draw the arms than the bent wire with square ends would be. In drawing the temper of a staff the other end may be clamped in a pin-vise, or a pair of brass lined pliers, which will keep that part cool and prevent its being colored; the copper wire being placed upon the staff, we heat it very quickly by blowing the flame with a blowpipe upon the outer end. As soon

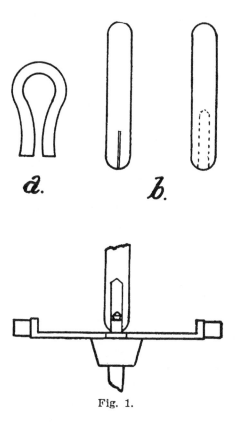

Fig. 1.

as the steel is drawn the desired amount, the copper wire is at once removed, which will prevent any further drawing. When this method is used there is no danger of affecting the balance in any manner. In order that our final work may be true, we must admit that the article to be pivoted must run perfectly true in the lathe; the chucks of our best American lathes are now well and accurately made. If we can clamp our work in them so it will run perfectly true, there can be no objection to their use, but it so often happens that it is almost true, and we are tempted to let it go instead of using cement, our only accurate way when the chucks will not hold our work properly. When a staff, or pinion, runs true in the lathe, the light reflected from the surface will appear perfectly still; but, should it not be true, the light re-

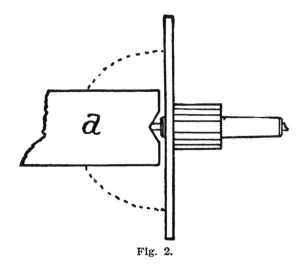

Fig. 2.

flected will waver. This is one of the most delicate ways of testing. The one important thing is, to have our work absolutely true, whether we use the chucks, or cement. There are many cases where nothing but cement can be used, for example: balance staffs, where hair-spring is underneath the balance, and those pinions where the wheel is very close to the end in both cases, only the pivot and a very small shoulder project beyond the wheel. In all cases of this kind when the pivot is broken on the opposite end, we must use cement, as there is no part to clamp a chuck upon. For work of this kind our cement chuck must have a very shallow center so the wheel will not touch the chuck when the pivot is resting at the bottom of the center as shown at *a*, Fig. 2, the dotted lines showing the cement. When the piece has been trued, and the cement is yet warm the lathe should be kept in motion until the cement sets; otherwise our work may settle and be out of true. After our work is perfectly true we are ready for the most difficult task, drilling the hole for our pivot; it being understood that we have already taken our measurements for the length of the staff when completed, and also its length before placing in the cement, that we may know just how long to make the pivot.

We come now to the most important part of the whole

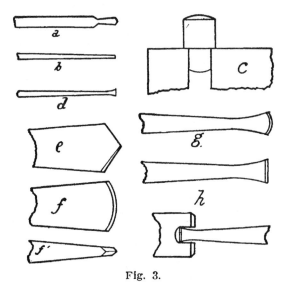

Fig. 3.

operation, making the drill. Most watchmakers think it is not worth while to make their drills as they can be bought so cheaply, which is true, so far as cheapness is concerned, but most of them are dear at any price, as they will not perform the work desired. My advice is to make all of your drills, if you wish to be successful with your pivoting, as you can soon learn to make a better drill than you can buy at any price. I would refer my readers to a former article on "Iron and Steel," as it will help them greatly in making the small drills. A thorough knowledge of the effect on the steel when hardened at high and low temperatures being very essential. A piece of steel so small needs some protection to prevent it from being burned while hardening. A small copper tube closed at one end, just large enough to hold the drill will protect it from contact with the flame and prevent it from being too quickly heated. Drills are made of many different shapes, each kind being used for some special work. We would not use the same kind of a drill for brass as we would for steel, neither would we use the same shaped one for hard and soft steel—the softer the metal the sharper the cutting angle and the harder the metal the more blunt will the cutting edge be.

In making our drills, we must select the very best grade of steel and when once familiar with a certain brand it is well to continue its use as another brand will not work the same.

Stubs, or crescent steel, is always good, and any of the high grade American drill rods can be depended upon. There are some of the best English needles that make splendid drills, as it requires a high grade of steel to make a good needle.

We can turn our drills down in the lathe like the ones we buy, a, Fig. 2, using about 1 m-m wire, hardening and grinding the sides flat; and the cutting edges on the end while hard. A better and easier way is to file them down tapering, slightly smaller than the finished drill is to be, using about .4 m-m wire as shown at b, Fig. 3, then by using a rounded surface stump in our bench block, as shown at c, same illustration; and resting the end of the wire upon it, and striking lightly with a hammer, the end will be enlarged and flattened, as shown at d; by making the drill larger at the end it will give better clearance and is less liable to break, which is often caused by the chips filling the hole around the drill, making it heat and break.

It is a good plan to harden the drills as soon as they are flattened, then by breaking off the point one can tell at once if they are hard, and the very point is the part that is liable to become burnt in hardening. The cutting edge may now be formed with the oilstone slip, making the shape of the drill like *e*, Fig. 2, if to be used for soft steel, or like *f*, if to be used on very hard steel. If the drills are tempered at all they should be only slightly, as it is necessary to have them very hard. It is a good idea to leave the points nearly full hard, tempering the back part of the blade to a blue; this gives strength to drill, and also a good hard cutting edge as shown at *g*. The drill should project out of the pin-vise but little more than the depth of the hole to be drilled, by doing so there is less danger of breaking the drills, and we can give

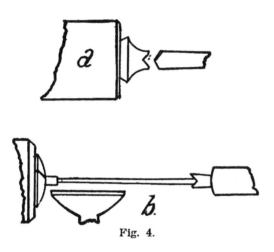

Fig. 4.

more pressure, which is quite important where the steel is hard. The pin-vise for all drilling should be light and strong and must clamp the drill firmly.

Many of the older workmen, who learned to do all of their work on the old bow lathe, think it impossible to do a hard job of pivoting on an American lathe. Let us see if we can find any reason for their argument; with the bow lathe the work is stationary and the drill revolves first forward, and then backward, caused by the motion of the bow. With the American lathe the work turns continuously in one direction; the drill is sharpened the same in both cases, round on the end; and sharpened on both sides as *f*, Fig. 3. To illustrate the difference more clearly, suppose we rub our hand over a piece of broadcloth or velvet in one direction. It appears very smooth, and continues so as long as it is rubbed in the same direction. But should we rub it forward and backward the surface will remain rough and cannot be smoothed. This illustrates the advantage of the bow-lathe

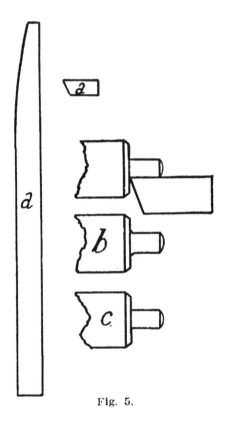

Fig. 5.

over the more modern ones, as the drill by reversing its motion keeps the surface rough and there is but little trouble about the drilling. While the American lathe turning constantly in one direction is liable to smooth the surface and also burnish it so the drill will not cut at all. We may overcome this difficulty in the modern lathes simply by running the lathe first forward and then backward, when we have all the advantages of the bow-lathe without its defects. If at any time while drilling the drill fails to cut, and the metal becomes burnished, we can only succeed in cutting through the burnished part by sharpening our drill nearly square on the end, as shown at *h*. Fig. 4, when, in most cases, it will easily remove the burnished part, when it can be sharpened again as at first.

Most workmen use too large drills, making the shell so thin that it is liable to crack when the plug is driven in to make the pivot. The drill to be used should be just a trifle larger than the hole in the jewel or plate that the finished pivot is to work in. The center in the staff, or pinion, should be small enough that the drill will cut it all out when the hole is drilled and the edges of the hole left perfectly square, as shown at *a*, Fig. 4. There should be no ring left around the pivot when finished.

It is necessary to be very careful in fitting the plug for making the pivot; the steel should be hardened and tempered to a blue and filed down, having but a slight taper, it should go to the bottom of the hole without being tight; by taking a trifle from the end it will go to the bottom and bind slightly; by taking off just a little more we know it will hold well when driven to the bottom of the hole, and it is not large enough to split the shell. In some cases it will be difficult to make the plug hold well, as the moment we begin to turn down the plug it will begin to work out, caused usually by its not being quite true. I have overcome this difficulty by making the plugs out of needles, drawing them to a blue, and after driving them in placing the other end in a female center in the tail stock, *b*, Fig. 4. This supports the wire until the pivot has been turned true and cut off to nearly the correct length, after which there will be no further trouble.

The method of grinding and polishing the pivots will be the same as that given in the article on staff-making. When replacing a square-shoulder pivot, it should be turned nearly down to size with a very sharp graver, the grinding being done with a bell metal grinder curved on one edge, and the corner filed often, the edge being kept sharp; otherwise the pivot will have the form shown at b, Fig. 5. It should be like c—perfectly square.

Very often the pivots of a hollow Swiss center pinion become badly worn, so much so that they would be too thin if ground smooth and polished. These can be nicely pivoted in the following manner:

First, cut off the pivot at the shoulder and drill a hole a trifle larger than the pivot is to be when finished. The plug is fitted the same as for other pivots; before driving it in we drill a hole in the center slightly larger than the center square so that the friction necessary to carry the hands will be on the inside of the pinion and not on the pivot we have replaced. When nicely polished no one would ever suspect that it had been pivoted.

THE BALANCE OR HAIR SPRING.

The Balance Spring, or as it is commonly called, the Hair Spring, is one of the most delicate parts of our time pieces, and upon its action depends much of the time keeping qualities.

There are several forms in use, the most common being the flat spring found in the cheaper movements, the Breguet spring (named after the inventor) used in the best movements, which is far superior to the flat, and the cylindrical spring used in marine chronometers and a few fine pocket watches. All of these springs are made from small wire, usually of steel, although for non-magnetic purposes, other metals have been used, palladium having given the best satisfaction for that purpose. Gold was tried, but nothing has yet been found that will give the satisfaction of steel.

A few words about the method of making will help us to understand the delicate spring better. The wire is first drawn through jeweled draw plates, which leaves the surface smooth and highly polished, quite an essential thing for a perfect spring.

A box made of copper, the inside of which is turned out a trifle larger than the finished spring is to be, is shown in Fig. 1; through the edge of the box are cut three or four openings, three being used for a close coiled spring, and four for a more open one. An arbor with two grooves is also shown in the same illustration. The wire for the springs is placed in position as shown in Fig. 2, two pieces being used, each long enough for two springs, the center of these pieces of wire being placed through the grooves of the arbor, then passing out through the openings through the edge of the box as clearly shown. A copper cap fitting loosely into the recess holds the spring flat, while the arbor is wound. When the space is completely filled, the top is securely fastened with coarse binding wire, and the springs are ready to harden. It will be readily seen that we will have four complete springs, and that each spring will have three other springs, filling the space between its coils; every fourth coil representing one spring, should only three

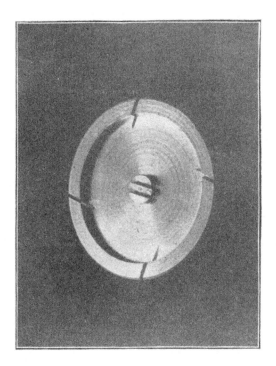

Fig. 1.

wires be used, then every third coil would be the same spring making the coils closer together.

In hardening the box containing the spring is covered with some substance that will prevent the oxygen of the air from reaching the polished steel while heating, as Potassium Cyanide; it is now heated to a cherry red and cooled in water. If we can prevent the oxygen of the air oxidizing the steel while heating, it will come out of the box as bright

as it was after drawing, and may be blued at once. Should
the surface become oxidized, then the springs must be care-

Fig. 2.

fully polished by hand, when they can be nicely blued on a
large flat piece of copper, which is evenly heated. We will
not find it necessary to make springs often, yet it is im-

portant to understand how it is done. We will now endeavor to explain duties they are expected to perform. Fig.
3 shows spring with collet and stud.

The length of a pendulum of a clock determines its
vibrations, so the strength of a balance spring will determine the vibraticns of the balance in a watch. We know
we can vary the vibrations of a pendulum by making it
longer or shorter, which is usually done by lowering or
raising the weight, as we lower the weight, making it
longer the clock will go slower as we raise the weight
making it shorter, the clock will go faster; with a watch or
any clock where the vibrations are controlled by a balance
and hair spring similar principles are involved. If we
lengthen the spring, our balance will vibrate slower, and if
we shorten the spring, the balance will vibrate more rapidly. We should also understand the effect of adding more
weight to the balance, or taking weight from it. When we
add more weight it has the same effect on the balance as
lengthening the pendulum, making the vibrations slower;
the opposite effect being produced when we remove weight
from the balance. Should we have a watch that is running
too fast and the spring cannot be lengthened, we can bring
it to time by adding another pair of screws, being particular
to have them of the same weight, otherwise the balance will
be thrown out of poise. Many of the best watches have four
screws placed at the quarters that are different from the
others, the heads being shorter and the threads longer and
fitting the holes more closely, having friction enough to hold
them in any position they are placed. These screws called
the timing screws, are used in bringing the watch to time.
Should it gain, the screws would be turned out, the weight
will be removed farther from the center, and will cause the
vibrations to be made more slowly; on the contrary, if the
screws are turned in, the weight will be brought nearer the
center causing the vibrations to be made in less time making the watch run faster.

Some balances have only two of these screws, one at
the end of each arm; if there are two or four, it is necessary to turn the two opposite ones the same amount, otherwise the balance would be thrown out of poise.

THE BALANCE OR HAIR SPRING.

With a pendulum, the force of gravity acting upon it, has a tendency to bring it to rest when once put in motion. The balance spring acts very much like the force of gravity, its action on the balance always bringing it to rest, so that a balance set in motion soon comes to rest if not acted upon by some other force that will keep it in motion.

The springs in common use are

a. The flat,

b. The Breguet and

c. The Cylindrical.

The flat spring is used in the cheaper movements, and is the most simple form.

The Breguet or overcoil spring is superior in many ways to the flat and are nearly always used in the better grades of watches. Some are made from flat springs by bending up the outside coil, then bending it down again, bringing the upper coil flat with the main part of the spring, the height of the overcoil being determined by the position of the stud in the balance cock. Some are hardened in form. The cylindrical spring is not often used in watches, although some of the finest ones are fitted with them. These springs are used in marine chronometers in preference to all others.

Nearly all watches we have to fit springs in have a fast train or 18,000 vibrations per hour, but as some of them have 16,200 and some 14,400 vibrations per hour, it is quite important that we understand how to find the vibrations the balance will make in a given time. The second hand makes one revolution every minute. It is carried by the fourth wheel which must make a revolution in the same time. We have a starting point now to work from. The teeth of the fourth wheel divided by the leaves of the escape pinion, will give the number of revolutions of the escape wheel per minute. Every tooth of the escape wheel,

95

gives impulse to each pallet, therefore, for one revolution of the escape wheel, there must be twice as many impulses as there are teeth in the wheel. The usual number of teeth is fifteen, so there would be thirty impulses or vibrations to each revolution of the escape wheel. Then in order to find the number of vibrations of any balance per minute, we would divide the number of teeth in the fourth wheel by the number of leaves in the escape pinion, and multiply this product by twice the number of teeth in the escape wheel, for example:

Fourth wheel has 80 teeth.

Escape pinion has 8 leaves.

Escape wheel has 15 teeth.

$$\frac{\overset{10}{\cancel{80}} \times 30}{\cancel{8}} = 300 \text{ vibrations per minute.}$$

or

Fourth wheel has 63. teeth.

Escape pinion has 7 leaves.

Escape wheel has 15 teeth.

$$\frac{\overset{9}{\cancel{63}} \times 30}{\cancel{7}} = 270 \text{ vibrations per minute.}$$

or

Fourth wheel has 80 teeth.

96

THE BALANCE OR HAIR SPRING.

Escape pinion has 10 leaves.

Escape wheel has 15 teeth.

$$\frac{\overset{8}{\cancel{80}} \times 30}{\cancel{10}} = 240 \text{ vibrations per minute.}$$

After determining the number of vibrations the balance must make in a given time (usually a minute) we are ready

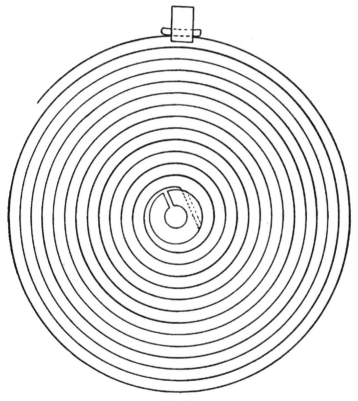

Fig. 3.

to select our spring; experience will aid us greatly in this task, as it is only by doing a certain piece of work often that we are able to do it quickly and well. It is so in fitting hair springs. There are numerous guages for testing the strength of the springs, also vibrators for timing them, but as the majority of workmen do not possess these they will not be considered at present, but the method given will be one that any ordinary workman may follow with good success, providing each step is carefully done.

It is well to become quite proficient with the flat springs before attempting the more difficult Breguet springs. It is best to buy the springs without collets, as the old collet is usually in good shape and fits, while a spring of the proper strength will seldom have a collet fitting the staff, if we buy them already colletted.

Many watchmakers think it a very difficult task to replace a hair spring, as it is such a delicate job. This is true to a certain extent, but it only requires carefulness, the same as with many other delicate operations the watchmaker is so often required to do, and I am sure if the following method is carried out, many who have dreaded the thought of "springing" will do it with ease and pleasure.

Our first step will be to select a spring that to the best of our judgment will be the one that will vibrate correctly. It may be too strong, it may be too weak; to determine that, we lay the spring upon the balance cock with the center of it exactly over the center of the hole jewel, and notice which coil comes over the inner regulator pin, as we cannot use a larger spring than this (we could use one slightly smaller, however). It is not necessary to collet the spring until we know the strength is correct. To count the vibrations, we place the spring in position on the balance, slip the collet on the staff and press it down until it holds the inner coil firmly upon the balance arm. We now clamp the coil of the spring that came over the inner regulator pin lightly in a pair of tweezers with fine points, and set the balance in motion, allowing the lower pivot of the staff to rest upon a smooth surface, a watch glass will answer the purpose nicely. The vibrations may now be counted for a minute. We only count each alternate vibration, or one-half of the number we found by calculation. The reason for this is, we count the vibrations each time the balance arm stops at a certain place, but

98

the spring has made two vibrations during that time; in other words, we only count as the balance turns in one direction, so we will take one-half of the number of vibrations we found the watch to have. For example, we have a fast train, 18,000 vibrations an hour or 300 per minute. We take one-half of 300, which is 150, the single vibration for one minute.

We count the spring for one minute, and find the vibrations to be too many. We know at once the spring is too strong, and cannot be used. The collet is now removed, and the spring may be replaced in the original paper unharmed.

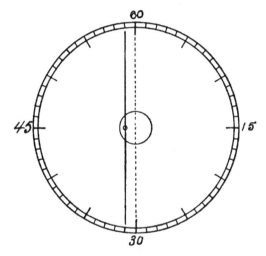

Fig. 4.

and a weaker one selected. After two or three trials, one should be able to find a spring of the proper strength. It will be readily seen that by this method the first time a spring is vibrated, it will be known whether it will do or not, and another can be quickly tried and counted.

A word about counting. I said count for one minute; many think half or a quarter minute will do as well; no, for several reasons. The second hand must make a revolution in one minute; it should go half way around the dial in one-half a minute, theoretically it would, but often in

99

actual practice, it does not. This will seem strange to those who have not noticed it carefully. To prove my statement the drawing (Fig. 3) has been made to represent the dial for the second hand; this dial is supposed to be correctly spaced, although many dials are not at all accurate, but we admit in this case that the dial is a perfect one, no cause for complaint from that source, but in fitting the dial to the movement the fourth pivot that carries the second-hand was not located exactly in the center of the dial. The two figures 60 and 30, and the dots representing them, are directly opposite; a line drawn from one to the other will divide the dial into two equal parts, but the pivot of the fourth wheel is at the left of this line as shown. A line is drawn through this pivot parallel to the one from 60 to 30, which shows the position of the second-hand when it has made one-half of a revolution. The drawing is somewhat exaggerated. To show it more plainly, the line of the second-hand passes through the 59th and 31st second; the result is

Fig. 5.

that during the first half of the revolution of the hand it would register 32 seconds, and during the last half it would register only 28 seconds, so if we should count our spring during the first half minute, and then count again during the second half, it would be impossible to get them the same, and would surely confuse us in determining the vibrations of our spring. To avoid this, I always count for one, two or three minutes. Another very important point about count-ing; it is only natural for one to begin the count when the second-hand reaches 60 or any other point on the dial, and end the count at the same point; by so doing we gain one count every time, as at the start we counted one before a vibration had been made, we must allow for this extra count either at the beginning or at the end; I think best to get the motion of the balance a few seconds before we are to begin actual counting, or what is still better, count one, two, three, four, etc., watching carefully which count comes at 60. Then begin with one again, as—one, two, three, four (60), one, two, three, etc., by so doing, we get the motion of the balance thoroughly in mind, and we are able to make a very accurate count, when we find the point where the vibrations are correct, we can make a slight bend in the spring to locate the place, and we are ready to fit our spring to the collet. This is not a difficult operation, but requires a steady nerve and care.

In most cases the center of the spring must be broken out to fit the collet, there should be enough broken out that when the collet is placed in the center, the space between it and the inside coil of the spring should be about the same as the space between two coils of the spring. The inner end should be bent at such an angle that it fits nicely into the hole in the collet, leaving the spring as true as possible when pinned in.

There are several methods of holding the collet, while pinning in the spring, only one will be given at present, which is quite an old one but works well in most cases. Photographs of a newer and still better device will follow in the next article. Most watchmakers slip the collet on a broach or a round file while pinning in the spring, this is not very satisfactory, but a piece of wire about 3 m-m in diameter, turned as shown in Fig. 4 does the work much better than the file or broach, and should be used for no other purpose. It is best made by hardening the wire and tempering to a blue, turning to shape afterwards; this insures it being true, while if it should be turned first and then hardened, it is liable to spring while hardening, thus rendering it unfit for use. The taper where the collet is held should be draw-filed until it is quite rough, this is important, as it prevents the collet from turning when the pin is being forced in.

The kind of brass wire used in making the pins is quite an important item. We should select the hardest brass we can get, as it should break before it will bend, most of the wire furnished by the material houses is altogether too soft; the best I have yet found is a good quality of ordinary brass pins which may be bought for a few cents and answer our purpose splendidly. They should be filed down to a long, slim taper.

To pin in the spring, slip the collet upon the colletter pressing it down firmly upon the taper, slip the end of the spring in the hole in the collet, being careful, of course, to have the spring enter from the proper side, then insert the pin from the underside of the spring, allowing the point to come through under the spring on the opposite side, keeping the spring as nearly level as possible while pinning it in. The pin is forced in as firmly as possible. Now bend the pin at right angles with the colletter (the pin being yet in the pin-vise) then by a twist of the pin-vise, it may be broken off just outside of the collet. Take a good, stiff pair of tweezers, placing one point on the broken end of the pin and the other point on the opposite side of the collet, and force the end in until it is even with the surface of the collet. The spring will now be perfectly solid, the projecting end of the pin may now be cut off with a sharp knife; it being on the underside of the spring, we raise the spring with the knife blade, and cut the end close to the collet. By cutting in this manner, the collet cannot slip, as the pressure forces it still more tightly on the taper.

The spring may now be bent so it is as true to the eye as possible in the flat and round, when it may be removed from the colletter and placed upon our balance wheel where the final truing should always be done in the calipers.

Before explaining how to true the spring, I will explain another method of pinning in the spring mentioned in the former pages. The taper which held the collet is liable to spread it, making it too large to fit the staff; it is necessary to force the collet on to the taper in order to have friction enough to prevent it from turning, so this method has the disadvantage of enlarging the hole in collet, which must be closed by bending; this is liable to damage the spring. To overcome this difficulty, the tools shown in Figs. 1 and 2 were made; both work on the same principle, clamping the collet on the top and bottom instead of depending on the friction as before. It will be readily seen that it is possible to hold the collet solid enough to force a pin in until the spring will be held very solidly without any danger of the collet turning or slipping. By using the one in Fig. 2 we can partly true the spring before taking from the colletter; in fact, we

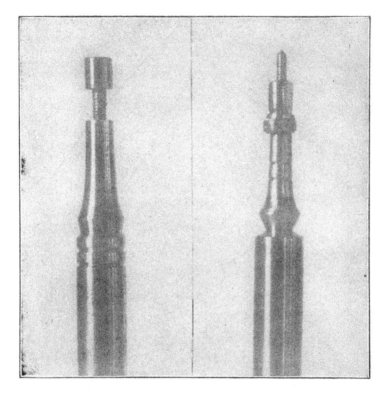

Fig. 1.   Fig. 2.

can do the most of it. The truing of the spring is a very difficult task to most workmen, but should not be so. The springs, as made, represent a perfect spiral, each coil gradually receding from the center; this being the case, if the spring is perfectly centered, placed on the staff and revolved

103

in the calipers, the coils would appear to move from the center to the outer end, or from the outer end to the center; it depending upon which direction we should turn the spring. Perhaps this may be better understood by watching a long screw as it revolves in the lathe; the threads appear to run from one end to the other; in fact, that is exactly what they

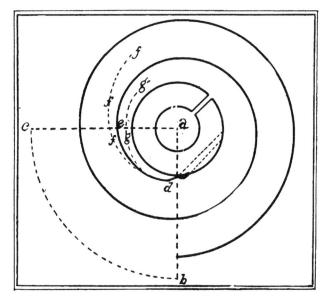

Fig. 3.

do, as the thread of the screw is a spiral wound around a cylinder; when our spring is true on the balance, it will have the same appearance as the screw threads in the lathe, only the coils will appear to run from the outside to the center, or from the center to the outside of the spring. As the coils of the spring are true when made, we must only bend the *inner* coil to do our truing, and only the first quarter coil of this should be bent, so in truing in the round, we

104

should not bend our spring beyond a quarter of a coil from the point where it is pinned in. This can be better understood from Fig. 3. The line *a. b.* passes through the point where the spring is pinned in, the line *a. c.* is one-fourth of a turn from it. Our bending must be done from the point where it leaves the collet *d.* to the point *c.* Should the spring

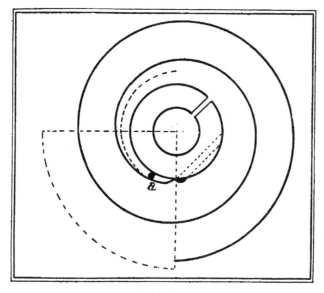

Fig. 4.

be out of center, as indicated by the dotted line *ff.*, then the spring should be bent about half way between *d.* and *e.*, which would throw the whole spring toward the right, bringing it nearer its correct position—shown by the heavy line. Should it be in the position shown by the dotted line *g. g.*, then it should be bent at the same point, but in the opposite direction, away from the collet. A few trials will

105

enable any one to true a spring easily. We should have a very fine pointed pair of tweezers fitted up for hair spring work and use them for nothing else. We will have no difficulty bending a spring away frcm the collet, but it i.. rather difficult at times to bend a spring towards the collet, as when we bend it that way, the elasticity of the spring

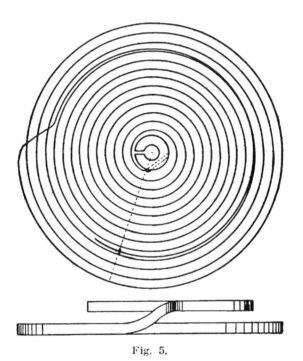

Fig. 5.

brings it back again to nearly the former position; to overcome this difficulty we may bend it as shown in Fig. 4. We wish to bend it at *a.*, but as we make the bend it immediately springs back again, we insert the fine point of a needle or file down a small piece of wire and insert as shown in the drawing. This acts as a fulcrum, and with our tweezers act-

ing just beyond that point, we are able to make the bend
without further trouble.

When our spring is not true, and we revolve it in the
calipers while it is on the staff, we will notice each time it

Fig. 6.

makes a revolution that one side seems to be thrown from
the center and the other side toward the center. We should
not try to watch each separate coil, but see the spring as a
whole. It would be better to liken the spring to a round
disc; if it were not revolving from the center it could be
easily noticed; even so with the spring, the trained eye can

as quickly see which side is thrown out, and it is only necessary to bend at the inner coil to bring it true.

It is an easy matter to true a spring in the flat, as we can usually raise it up on the low side, or bend it down on the high side by a gentle pressure with the tweezers. There are times, however, when the spring seems a trifle stubborn, and we must bend it by taking the inside coil in our fine pointed tweezers and twisting enough to bring it flat near the point where it is pinned in.

When our spring is true in the flat and round on the balance it is ready to pin in the stud, and all the rest of the truing necessary to bring the spring perfectly flat can be done by bending the outer coil at or near the stud after pinning it in the stud as flat as possible. The regulator should be placed on the slow side, and the outside coil of the spring so bent that it will vibrate between the pins in the regulator; then move the regulator toward the fast side, bending the spring just ahead of the pins until the outer coil of the spring will vibrate between the pins the full sweep of the regulator; when this has been accomplished, we should see that the coils of the spring (when the balance is at rest) are the same distance apart on all sides; should they be close on one side and far apart on the other, we should correct them by bending the outer coil just beyond the regulator pins. The spring should never be "cup shaped," which is caused by the stud being too high, or the hole in the collet too low. The hole in the stud should be exactly the same height as the hole in the collet, otherwise the spring cannot lie flat in the watch; theoretically, when the spring is pinned in the collet and stud, and the balance is at rest in the watch, there should be no strain upon the spring in any direction, as every part is in a state of rest. This being so, the spring must be flat, and the coils on all sides must be the same distance apart.

It may be well to state here the effect the regulator pins have upon the timing of a watch, as the rate may be varied much more than most workmen think by them. If we make the space between the pins so the spring is just free, having no play, then the spring will stop its vibrations at that point, and will run faster; now open the pins until the

108

spring may vibrate freely between them and the watch will begin to lose. Again, while the pins are far enough apart to allow the spring to vibrate freely between them, bend the spring so it will rest against one pin when the balance is at rest, but will also vibrate between the pins when the balance makes it full vibration; in this case the watch will lose when making its long vibrations or when first wound up, and gain during the short vibrations or when it is nearly run down. These little things change the rate of a watch so greatly they should be well understoood; to illustrate this more fully, an incident in my own experience may be mentioned: A gentleman who owned a very fine watch came into the store and stated that the regulator was not in the center on his watch, and asked if I could make the watch keep time and have the regulator where it should be. After examining the curb pins, and seeing they were far apart and the regulator on the fast side, I told him it could be easily done. I bent the pins closely together, barely allowing the spring to vibrate between them; then, while he was watching, brought the regulator to the center, moving toward the slow side nearly half the width of the balance cock. I handed him his watch, stating that if he would come in the next day, the watch would in all probability be a little fast; he looked at me in a way that said very plainly, "I don't believe it," but went on his way, and the following day came in and compared with the chronometer, and to his surprise he was nearly two minutes fast. The balance of the regulating was done by gradually opening the curb pins until the watch was brought to time, and the regulator still remained in the center. It is needless to say, from that day forward, I had another friend. When the watch was nearly regulated, I explained the principle very carefully to him, and he fully appreciated it.

A few years ago only the very highest grade movements had the Breguet or overcoil springs, but today nearly every movement has them. They are much more difficult to handle than the flat springs, but if the latter has been mastered, the former may soon be.

Fig. 5 shows a Breguet spring as seen from the top and also from the side, showing the elbow or double bend. There are a great many forms of these springs, but the principle is the same in them all. A larger number of coils are used than in the flat, and a greater latitude is allowable in selecting our springs, the regulator pins being nearer the center than the outside of the spring. The overcoil must be brought toward the center in order to pass through the pins and vibrate between them. After selecting a spring of the proper strength and pinning in the collet, by placing the center of the collet over the balance jewel, we notice which coil of the spring comes over the regulator pins, and that is the coil the overcoil should follow. We usually say the overcoil should follow this coil, but it should not, as can be readily seen by looking at Fig. 5. The overcoil does not follow one of the coils of the spring, but is the same distance from the center at all points, while the coils of the spring being a spiral, are not the same distance at any two points, hence the spring could not follow a coil, and also follow the sweep of the regulator, which it must do in order to vibrate between the pins. The height of the overcoil is determined by the distance the hole in the stud is above the hole in the collet. Should the overcoil be too high the spring will be low at the outside, and should it be too low, the spring will be high at the outer edge. What was said of the flat spring is also true of the Breguet when it is pinned in and true in the flat and round, there should be no strain on the spring at any point, but it should be in a perfect state of equilibrium.

The Breguet spring is superior to the flat one in many ways, the most important point being the absence of side pressure on the pivots. On account of the overcoil being pinned in or near the center of the spring the coils open out more evenly, and there is but a very small amount of side pressure upon the pivot, which is very helpful to the close timing of a watch. The spring should be pinned in at about equal turns as shown in Fig. 5 by the dotted line. The point where the spring enters the collet and the point passing through the curb pins are in line. In order to make the spring isochronal we must at times vary this rule by pinning in some cases a trifle less and in others a trifle more than equal turns in order that the long and short vibrations

may be made in the same time. More will be said about this
when we get to adjusting.

The overcoils are made in a great many different forms,
each having a different effect on the timing.

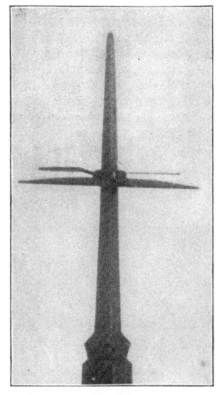

Fig. 7.

Fig. 6 is a photograph of a Breguet spring showing the
overcoil very plainly, and Fig. 7 is a side view of the same
spring, showing the elbow; it will be noticed that the spring
is not flat, the outer coils being lower than the center. This

111

is caused by the weight of the spring, and should always be considered when timing a watch in different positions, as in many cases, particularly with the Breguet springs, the coils will rub on the end of the stud in one position, while in the opposite position they seem to, and do, clear the stud easily; this is the cause of much irregular running. We should also be very careful that the second coil in a flat spring does not strike the inner regulator pin, particularly when the balance is making full vibrations, as this has a tendency to make the watch gain time.

Much trouble is experienced by two coils catching in the regulator pins. This can be overcome by making the curb pins just long enough to reach nearly to the bottom of the spring, and allow but little space between them. Should another coil, by a sudden jar, be thrown over the pins, it will at once resume its normal position.

THE LEVER ESCAPEMENT.

There has been so much written about the Lever Escapement in the past that there seems to be no field for new thought at the present time, but when we realize that over ninety per cent of the watches in use to-day have the lever escapement, and that it seems to puzzle many watchmakers to locate difficulties, one may be pardoned for trying to add a little to what has already been said on the subject. It will be the aim of the writer to present practical points that will greatly help the beginner, and possibly improve the methods of the more advanced workman. These points will be illustrated in such a way that they can be clearly seen, by drawings, and by photographs of actual escapements, showing their good and bad points.

In order to be able to understand the escapement, we must know the names of the different parts, and what duties they are expected to perform. Some parts are known by several names, and we should be familiar with them all. For the present, we will only speak of the escape wheel and pallets. We have two kinds of escape wheels, those used in the English lever watches called the pointed tooth, where all of the impulse is on the pallets, and the kind used in nearly all watches to-day, the club tooth, where the impulse is divided, part being on the teeth of the wheel, and the remainder on the pallets. In the former the points of the teeth are liable to become bent, which would greatly interfere with the running of the watch. The club tooth is stronger and less liable to damage while in the hands of a careless workman.

The impulse or lift is given by the *inclined planes* of the escape wheel teeth acting against the inclined planes of the pallet stones in the club-tooth escapement, and the point of the tooth acting against and passing over the inclined plane of the pallet stone in the pointed tooth escapement. By the

113

drawing, Fig. 1, we are able to better understand the principle and the manner in which the impulse is imparted from
the tooth of the wheel to the pallet stone. Let W represent
a wedge, which is a form of an inclined plane, resting on the
block A. At B a piece resting upon the wedge W, but free
to move in the direction of the arrow. By moving the wedge
to the right as indicated by the arrow, the piece B will
move upwards as shown.

In an escapement we have similar principles, but in no
case does any part move in a straight line. The pallets move

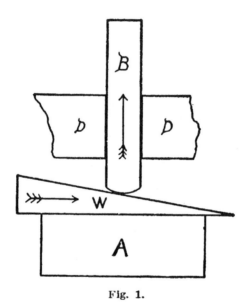

Fig. 1.

in a circular direction, having for their center the pallet
arbor. The teeth of the wheel also move in a circular direction, having for their center the escape pinion. All these
points move in a circular direction and many complications
arise which makes the escapement hard to understand, and
is liable to confuse one. For instance, it is rather difficult

114

to understand that if we move one pallet stone *out*, that it will increase the lock on *both* pallet stones as much as we set the one out, or should we set either stone back, the locking on each will be decreased exactly the same amount. This will be gone into more in detail a little later.

The pallet arbor being the center from which the pallets rotate, we can see as one pallet stone is moving out, the other must be moving in. As it will be necessary for many of our drawings and actions to be shown by degrees, we should all understand what is meant by a degree. We know that all circles have the same number of degrees (360), but only circles of the same diameters have degrees of the same size. A degree is one of the 360 equal parts of the circumference of a circle. A right angle contains 90 degrees or one-fourth of a circle.

If one end of our pallet arm should move outward one degree, then the other end must move in one degree, even if the two arms should not be the same length. For this reason one of the pallet stones has a greater angle than the other. One pallet arm being longer than the other the angle will be greater as the degrees on a circle of that size are larger than those of a circle the size of shorter arm.

Fig. 2 shows the pallets used in most lever escapements. The pallet steel or pallets, as they are commonly called is shown at A, the pallet stones, R. and L., the locking faces (*a*) and the impulse faces (*b*).

We have many different names for the pallet stones, for short we say the "R." and "L.," meaning the receiving and let off, also called the engaging and discharging. We will speak of them as the R. and L., or receiving and let off. The tooth of the wheel locks against the locking face of the stone (*a*), while the balance makes its vibration. As the tooth of the wheel slides across the face (*b*) it gives the balance its impulse.

The amount of locking the pallets have is a very important thing for the good performance of a watch. If it is too little the teeth are liable to fall upon the impulse face, a very serious defect; on the other hand, if the locking is excessive, then too much power is required in unlocking, and all power thus used is just that much less for the impulse

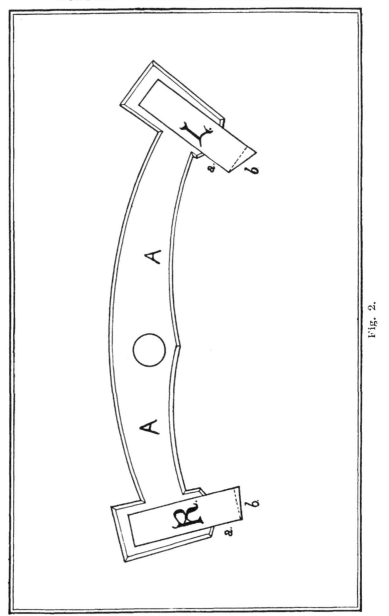

Fig. 2.

to the balance. We should have the proper amount of locking in our mind, so that the moment we glance at the lock
on a pallet stone we can instantly tell if it is correct or not.

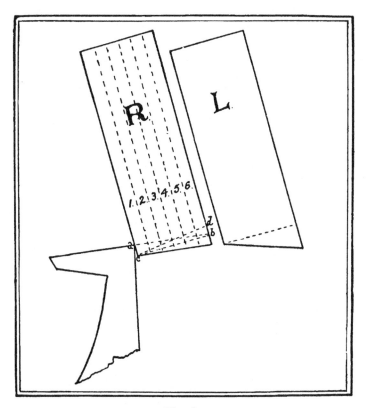

Fig. 3.

In order to aid us in getting this point thoroughly impressed
upon our mind, the drawing shown in Fig. 3 has been made.
The R. stone is divided into six equal parts shown by the
dotted lines lengthwise. The amount of the lock should be

117

equal to the width of one of these spaces, or in other words, the amount of lock should be about one-sixth the width of the impulse face of the pallet stones in the club-tooth escapement. In Fig. 3 a tooth is shown locking the proper

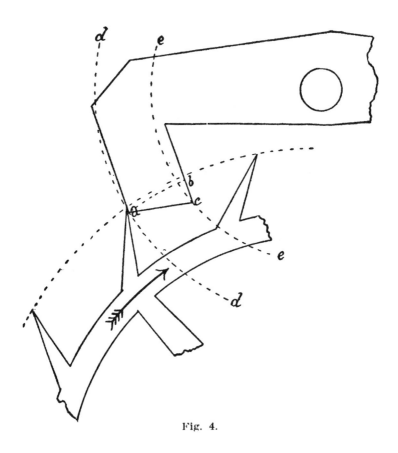

Fig. 4.

amount. The locking is denoted also by the dotted line crossing the stone (a-b). The L. stone is also shown by the side of the R. stone. Its greater angle is clearly seen, but we will learn later that each of them have the same number of *degrees* of impulse or lift. The real impulse is greater

118

than it appears on account of the circular movement of the pallets and wheel, for instance the inclined plane would seem to be that shown by drawing a line at right angles to the locking face as (*b-c*), but in reality, the inclined plane would go to the line *c. d.* If the R. stone should have a *perfectly square* face, there would still be some impulse. This can be better shown in Fig. 4, where all of the lift or impulse is on the pallet stone. The wheel moves in the direction the arrow indicates; the dotted line shows the path of the points of the teeth, the point of the tooth is locking against the pallet stone the proper amount, it will continue to lock until the corner of the pallet stone is raised above the point of the tooth when the tooth at once acts upon the impulse face of the stone. The dotted line *a. b.* represents the surface the stone would have and allow the tooth to pass over without giving it any impulse, so the triangle *a. b. c.* represents the wedge or inclined plane of the pallet stone. The locking face of the stone is set at such an angle that the wheel must recoil slightly in unlocking. This angle keeps the lever against the banking pins, and also prevents the guard pin from coming in contact with the roller. When this takes place, we say, the pallet stone has "draw" or "draft." The locking corner of the pallet stone is always on the dotted circle *d. d.* and the let-off corner is always the same distance from the center of the arbor as shown in the drawing, and cannot change when the pallets are moved in different positions. Fig. 5 shows this point in a very clear manner. A is the center of the escape wheel, P the center of the pallets or pallet arbor. The larger dotted circle represents the circumference of the wheel, and the smaller one the path of the locking corners of the pallet stones. They must always be the same distance from the center of the pallet arbor. If all points on the locking face of pallet stones were the same distance from pallet arbor, we would have a "dead beat" escapement, i. e., one which has no recoil in unlocking. Some of the first escapements were made of this form, such a locking face is denoted by the solid curved line *d. g.* and *d'. g'.* We must bear in mind that in most cases the R. and L. stones act entirely different; for instance, the locking face of the R. stone is shown by the line

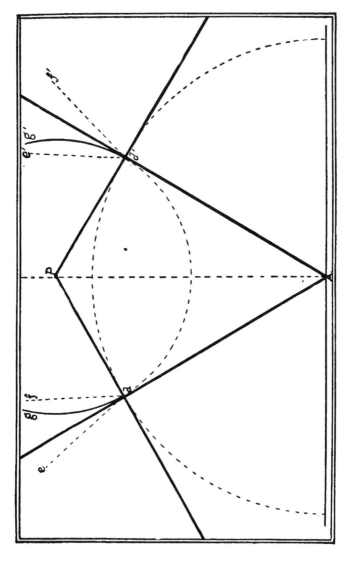

Fig. 5.

120

d. f. and is *inside* of the circle. For the L. stone the line *d'. f'.* shows the locking face and is *outside* of the circle. This is a very important point, and all should study it until it is well understood. Should the locking faces of our pallet stones be set on the lines *d. e.* and *d'. e'.*, we would have no draw, but would give impulse instead; in fact, it would have impulse if set *outside* of the circle on the R. stone or *inside* of the circle on the L. stone. If they are set on the circle they would have *neither* draw nor impulse, being a "dead beat," and if the R. stone is set *inside* of the circle, and the L. stone *outside* of the circle, we *will* have draw. The greater the angle on which the stones are set, the greater the amount of draw to the pallet stones. There should be just draw enough to hold the lever against the banking pins, and if we bring the lever away from the pins, but not enough to unlock, there should be sufficient draw to immediately bring the lever back to the banking pin again. If the draw is excessive, then it will take too much force to unlock the pallets, and such loss of power must reduce the motion of the balance. When the escapement is analyzed more fully, these points will be understood without any difficulty.

In the drawing (Fig. 5) the heavy lines show the foundation on which all lever escapements are built, the opening of the pallets 60° is found by laying off one line 30° to the right, and the other 30° to the left; the other two lines are at right angles to them at the circumference of the wheel, and where they cross the center line locates the pallet arbor. This should be clear before attempting to draw an escapement. There is no way of learning the principles of an escapement, and no way of impressing it upon the mind thoroughly like drawing one correctly and *knowing* why the lines are drawn at certain angles, etc. This will be one of the features of these articles on the "Lever Escapement," and my advice to all who can possibly do so, is to make the drawing when such are explained, even if in only a crude manner, as it will bring out new points never before noticed.

Fig. 6 shows a club-tooth locking the correct amount at *a.* If this drawing is kept in mind, one can tell the moment the escapement is examined whether the locking is correct or not. At *b.* is seen the pallet stone after it has moved up-

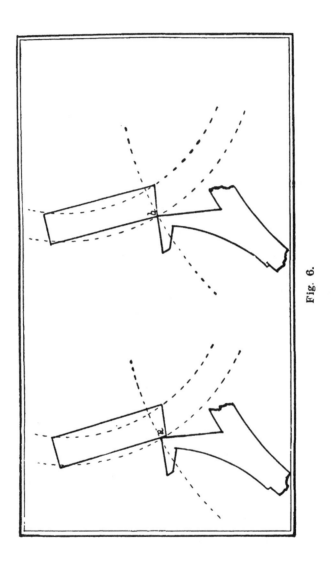

Fig. 6.

wards enough to unlock the tooth, and it is just ready to give "impulse" or "lift." As the tooth moves forward, the pallet is forced upward by the combined action of the impulse face of the tooth and that of the pallet stone. When the back of the tooth passes off from the let-off corner of the stone the wheel is perfectly free for a moment, and the "drop" occurs, this drop should be the same when a tooth leaves the "R." stone as when one leaves the "L." stone, otherwise we would have unequal drop, and a watch would sound as though it were out of beat. When a tooth leaves the "R." stone we have the *inside* drop, and when a tooth leaves the "L." stone, we have the *outside* drop; should this be the greatest, it may be corrected by moving the pallet stones farther apart. If the inside drop is the greatest, then it would be corrected by bringing the stones closer together. A point here should be well understood; in either of the above cases, we must move the stone only one-half as much as we desire to change the drop. If we desire to decrease the inside drop, and we bring the stones closer together, say 1-10 of a millimeter, we have *decreased* the inside drop that amount, but we have also *increased* the outside drop the same amount, making a difference of 2-10 of a millimeter. From this we learn to only move the stones one-half as much as we desire to change the drop.

Another point that is hard for some to understand, and will be repeated frequently to impress it on our mind, is this: If one pallet stone, is set out or toward the wheel, or away from it, it will increase or decrease the locking on *both* pallet stones just as much as *either one* is moved. Should the locking be too light and one stone is set out, it would increase the locking enough, perhaps, but if the watch was in line at first, it would be thrown out of line by moving only one stone, so it would be necessary in order to keep our watch in line to set each stone out *one-half* as much as we desire to increase the lock. Some think the locking may be increased by opening the banking pins. This is not so, as the locking is determined by the position of the pallet stones. By opening the banking pins, the locking appears greater as the "draw" causes the "run," the run is the movement of the pallets *after* the lock. If the watch is banked to the drop, there will be no run, but as the banking pins are opened, the lever is allowed greater angular motion, and the draw of the pallet stones holds the lever against the banking pins. Some of these terms are a trifle confusing, but with the series of photographs which will follow, these difficulties should disappear.

DRAWING THE ESCAPE WHEEL AND PALLETS.

It will be necessary for us to have a few drawing instruments, and understand their use before we can attempt to draw any part of the escapement. A very good set of instruments may be bought for a few dollars. We can get along for the present, however, with a drawing-board a trifle larger than our drawings are to be when completed; 18x24 inches will be a convenient size. This board should have the ends and sides right angles to each other, or at least the bottom and left-hand side should be so. A T-square with the blade as long as the board, a pair of dividers, a combined pen and pencil dividers, two triangles, one 45°-90° or right angle, the other 30°-60°-90°; a good protractor divided into ½ degrees; a set of thumbtacks and some good drawing paper. With the tools mentioned we will be able to make good pencil drawings. Then if we wish to ink them in, we will require a bottle of Higgins' India ink and a bow pen.

The triangles and protractor are more convenient if made of celluloid, as they are transparent and we can use them to better advantages. We can get a triangle (right angle) and protractor combined, which is very satisfactory.

Figure 7 shows the drawing board A., and the various instruments one should have to start; the better the quality of the instruments the better the work may be expected, but to begin with, cheap ones may be used, although they are not so accurate. On the drawing board is shown the following:

B. T square.

C. Right angle, triangle and **protr**actor, combined.

D. 30°-60°-90° triangle.

E. Dividers.

F. Pen and pencil dividers.

G. Hard pencil.

H. Pencil eraser.

I. Paper with thumbtack in each corner.

J. Metric Ruler.

It will be necessary for us to understand the geometrical terms used in order to follow the drawing intelligently. In order that the reader may become more familiar with the terms, they are defined as follows:

1. A *point* has neither length, breadth or thickness, and shows position only.

2. A *line* has but one dimension—length.

3. A *straight line* is one that does not change its direction throughout its length.

4. A *curved line* changes its direction at every point.

5. A *broken line* is made of several straight lines, having different directions.

6. *Parallel lines* are the same distance from each other at all points.

7. A line is *perpendicular* to another when the angles formed by it are right angles.

8. An *angle* is the opening between two lines that meet; the vertex is the point where they meet.

9. If a perpendicular line is drawn from the center of a straight line, two *right angles* will be formed, each containing 90°.

10. An *acute* angle is less than a right angle.

11. An *obtuse* angle is greater than a right angle.

12. A *surface* has only two dimensions, *length* and *breadth*.

13. A *polygon* is a figure having three or more sides. A polygon having three sides is called a *triangle;* one of four sides a *quadrilateral;* one of five sides a *pentagon;* one of six sides a *hexagon;* one of eight sides an *octagon*, etc.

14. A *circle* is a plain figure, bounded by a curved line, called the circumference, every part of which is equally distant from a point within, called the center.

15. The *diameter* of a circle is a straight line passing through the center, terminating at the circumference.

16. The *radius* of a circle is a straight line drawn from the center to the circumference, and is equal to one-half of the diameter.

17. An *arc* of a circle is any part of its circumference.

18. A *chord* is a straight line joining the extremities of an arc.

Fig. 7.

THE LEVER ESCAPEMENT.

19. An angle is measured by the number of degrees contained between the two straight lines which form it.

20. A *tangent* to a circle is a straight line which touches the circle at only one point, and is always at right angles to a radius drawn to that point.

21. When two or more circles have the same center or are drawn from the same point, they are called *concentric circles*.

22. A *degree* is one of the 360 equal divisions of a circle. A half circle contains 180 degrees; a quarter circle, 90 degrees, etc. The sizes of the degrees vary according to sizes of the circles.

23. A *protractor* is an instrument divided into degrees and half degrees, used for measuring or drawing angles.

The protractor and the degrees it is divided into must be well understood before we attempt to make our drawings. Every circle no matter how large or how small has 360 degrees; in other words, a degree is the 1-360 part of a circle. The degrees of a large circle will be large and those of small circles must be small. Degrees do not denote *size* but *parts*. A complete circle then has 360°, a half circle 360 ÷ 2 = 180°, one-fourth of a circle 360 ÷ 4 = 90°, etc. A protractor represents usually a half-circle, divided into 180 equal parts representing degrees, and in most cases these are subdivided, giving ½ degrees; the center of the protractor should always be placed at the point from which we intend to lay off the degrees; to illustrate (Fig. 8) we have the horizontal line a-b. and the line perpendicular to it e-d. we wish to lay off 30° to the right and 30° to the left of the line c-d.; from point c. we place the protractor with its center at c.; place a pencil dot at the 30° mark at the right, and the same on the left, then by drawing a straight line from c. through each of the dots, we have two lines, one 30° at the right of the line c-d., and the other 30° to the left or 60° from each other. No matter what the size of the circle that should be drawn, if these lines were continued our angles would be the same. For this reason, we are able to make a large drawing of an escapement or any part of it, ten, twenty or thirty times the actual size, and then reduce it to actual size very accurately.

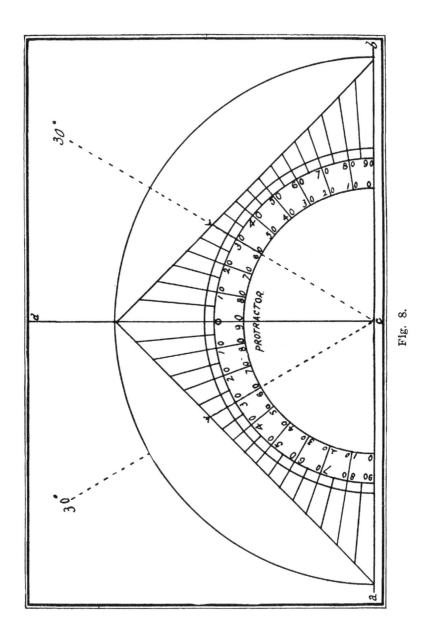

Fig. 8.

THE LEVER ESCAPEMENT.

All club-tooth escape wheels in use at the present time have fifteen teeth; the space between two teeth must contain as many degrees as $360 \div 15 = 24°$; this we should keep in mind. If it were possible to make an escapement absolutely perfect, having no recoil, we could make the width of the pallet stone and tooth one-half of the space between two teeth $24° \div 2° = 12°$, but as we must allow side shake of the pivots in the jewels, and we must have draw to the pallet stones in order to keep the guard pin from rubbing on the roller, and when we have draw, we must have recoil to the wheel in unlocking, we find it impossible to make our pallet stone and tooth more than eleven degrees wide where all parts are nicely constructed, and $10\frac{1}{2}°$ is what is used in most cases. This will be the amount in these drawings. The difference between $12°$ and the $10\frac{1}{2}°$ (the width of tooth and pallet stone) is $1\frac{1}{2}°$, which is the amount of drop to the escape wheel.

The pallets are free to move with the pallet arbor, and from the time of lock to the drop, the pallets should move just $10°$; of this $1\frac{1}{2}°$ represents the locking, $4°$ the impulse on the pallets and $4\frac{1}{2}°$ the impulse on the tooth of the escape wheel of the $10\frac{1}{2}°$ we have for the width of the pallet stone and escape wheel tooth, $5°$ represents the pallet stone and $5\frac{1}{2}°$ the tooth. It will make it much easier to remember, if we keep in mind that whole numbers ($5°$-$4°$) will always be on the pallets, and the *fractional* numbers $5\frac{1}{2}°$ and $4\frac{1}{2}°$ are always on the teeth of the wheel; by keeping them thus associated, we will have but little trouble in remembering.

We may now proceed to make our drawing of the escape wheel and pallets (the fork and roller action will be taken up later). Our paper is placed on the board as smoothly as possible, and fastened with the thumbtacks; place the T square against the left edge of the board and draw the base line A. B. from left to right near the bottom edge of the paper. Next bisect this line and draw a perpendicular line to it C. D. If the drawing board is perfectly square, this may be done with the T square from lower side of the board, but as most boards are not very accurate, the better way will be to use a right angle triangle, resting it on the T square,

129

which is still in the same position as when the line A. B. was drawn. This perpendicular line passes through the center of the pallet arbor and starts from the center of the escape wheel (only half of which will be drawn); the primitive diameter of which is denoted by the semi-circle W. W. W. The diameter of this circle will be determined by the size of our paper, the opening of the pallets (the distance from the locking face of one pallet stone to the locking face of the other one is 60°) may be found by laying off 30° to the left, and 30° to the right as in figure 9, giving us lines C. E. and C. F.

Our next step will be to locate the pallet arbor. This is done by drawing tangents to the circle W. at the points where the 30° lines cross it, or in other words, it will be a right angle to the lines C. E. or C. F. at the point where they cross the circle W.; where these two lines cross the line C. D. will locate the pallet arbor P. We draw the lines P. G. and P. H. We can now draw our pallets locking on the R. stone, at rest, or locking on the L. stone. First, we will draw them locking on the R. stone, the circle drawn, W. passes through the locking corners of all of the teeth, the impulse faces of the teeth then must be outside of this circle, and the locking, and the impulse of the pallet must be inside of this circle, the impulse face of the tooth will have 4½° lift. The lock will be 1½° and the impulse of the pallet stone will be 4° (4½° + 1½° + 4° = 10°, total movement of pallets); place the center of the protractor at P. (center of pallets) and mark 4½° above line P. G., 1½° below same line and 4° below the 1½° or 5½° (4° +1½°) below the line P. G. Draw straight lines from point P. through these points giving us lines P.-1, P.-2, P.-3. For the width of our pallet stones and escape wheel teeth, we lay off from the center of the wheel to the right of the line C. E. 5° for pallets and to the left of same line 5½° for width of tooth, giving lines C. 4, C. 5. We also draw a line 5° at the right of line C. F. for the width of the L. stone C. 6. From center of pallets P. draw an arc of a circle a-a passing through the points where 30° lines cross the circle W.; this arc of a circle shows the path of the locking corner of each pallet stone, they are always on this circle at some point, we also draw arcs through the points where the 5°

130

lines cross the circle W., from center of the pallets. The arc on the left b-b represents the path of the let off point of the R. stone and the arc on the right c-c shows the path of the let off point of the L. stone. These are important points, as will be seen when we draw in the impulse faces of our pallet stones.

To get the full diameter of our wheel, draw a circle from center of the wheel through the point where the 4½° line crosses the circle that shows the path of the let off point of the R. stone. This circle will be concentric with W. The back part of the tooth will be on this circle.

When a tooth is locking on the R. stone, the let off point of the L. stone must be at the circumference of the wheel, the let off point of this stone is always on the arc of the circle (c-c), then it must now be at the point where these two circles cross each other, and through that point we draw the line marked A, and above this line lay off 4° for our impulse to the L. stone.

We can now draw in the impulse face of the tooth and the impulse faces of our pallet stones. First, we will draw in that of the tooth. We have 5½° for the width and 4½° for the height of the impulse face of the tooth, as shown by the lines drawn. The back of the teeth are on the outer circle, and the front or locking corner are on the inner circle, so if we draw a straight line from the point where the line C-5 crosses the outer circle, to the point where the line C. E. crosses the inner circle, we will have the impulse face of our tooth. In drawing that of the pallet stones, we know the locking corner is always on the circle a. a., and the let off point of the R. stone is on the circle b. b. The locking corner is also on the line P. 2, and the let off point on the line P. 3, so the impulse face of this stone would be drawn from the point where the line P. 2 crosses the circle a. a. to the point where the line P. 3 crosses the circle b. b. The L. stone is similar, the locking corner being at the point where the line p. 7 crosses the circle a-a., and the let off point where the circle c-c. crosses the line P. A., connecting these two points with a straight line we have the impulse face of the L. stone. We should now observe that neither the locking nor let off points are on the straight lines 5° apart that were

131

laid off for their width. This is very plainly seen at the let off point of the R. stone, as it is on the circle b. b. and 5° line C. 4 is some distance from it. This is caused by the circular movement of the pallets. If you follow the arc of the circle, it will be readily seen that when the pallet has been raised until the let off point is at the circle w. w. or primitive diameter of the wheel, then the circle and the straight line both pass through the same point; the farther we move our pallets the greater the distance will be.

It is necessary that the front of the tooth be cut away in order that the tooth may come in contact with the pallet stone only at one point. In order to give the required clearance, we lay off 24° at the left of the line C. E., placing the center of the protractor at the locking corner of the tooth, giving us line d-d. We now draw a circle (e-e) to which this line is tangent. The front faces of all of the teeth will be tangents to this circle. We may now proceed to locate the other teeth. If we had a complete circle we could step it off with the dividers into fifteen equal parts, but as we only have half of the circle, this can not be done. We could lay off 24° from the locking corner of our tooth, and get it fairly correct, but we have a more accurate method than that, we know the line C. D. forms a right angle with the base line A. B. The line C. E. which passes through the corner of the tooth is 30° from the perpendicular line; in all we have the sum of 90° and 30° or 120°. This divided by 24° (the number between two teeth) 120 ÷ 24 = 5, there must be exactly five spaces from the tooth we now have to the base line on the right. We can set our dividers so they will exactly step it off in five spaces, each of these points will be the locking corner of a tooth. The other teeth may now be located by the dividers. The impulse faces of the other teeth may be drawn the same as the one that is finished, the back portion of the teeth may be made of any desired form, but must be cut out enough to allow for the recoil in unlocking. The rest of the wheel has no definite size and much depends upon the maker's individual taste, but it is important that the front face and the impulse face should be made as nearly correct as it is possible to make them.

The drawing of the locking faces of the pallet stones

have been purposely left until the last. It might be well to look over Fig. 5 of the last article before we proceed with the drawings, as a point here differs with nearly all drawings. By this method we can draw our pallets in any position, either locking on the R. or L. stone, or at any intermediate point, and our draft will always be the same, which cannot be said of all methods.

We now draw a line perpendicular or right angle to the $1\frac{1}{2}°$ line (P. 2) from the locking corner of the R. stone f; at the right of this line we lay off either 12° or 15° for the draw. We will use 12° on each stone. This line gives us the locking face of the R. stone. We also see that it is inside of the circle a.a., which represents the path of the locking corners of the pallet stones.

For the L. stone we proceed in a similar manner; in all cases we draw first a perpendicular line from the line passing from the center of the pallet arbor through the locking corner of the stone, the perpendicular line always starting from the locking corner of the stone. Then at the right of this line lay off 12° for draw. When these lines are drawn for the L. stone, we notice the locking face is outside of the circle a. a.

We have now the locking and impulse faces of both pallet stones drawn. The other faces may be drawn parallel to the locking faces, and the pallet steel drawn in. This, as in many other cases, is a matter where one may please himself, the acting faces only requiring perfect form.

When the wheel and pallets have been drawn in with the pencil, those who desire may ink them in, using the bow-pen in the compasses for the circles, and the hand bow-pen for the straight lines, being careful to have the pen lean away from the ruler, otherwise it is liable to blot the paper.

After the wheel and pallets are drawn, we are ready to take up the fork and roller action. The length of the fork may vary with reference to the diameter of the escape wheel, being from .5 to .7 of its diameter, and in some cases even greater, in fact some Swiss watches which are made with a very showy escapement, have been found where the length of the lever was one and one-half times the diameter of the escape wheel; the proportion of the fork and roller would still

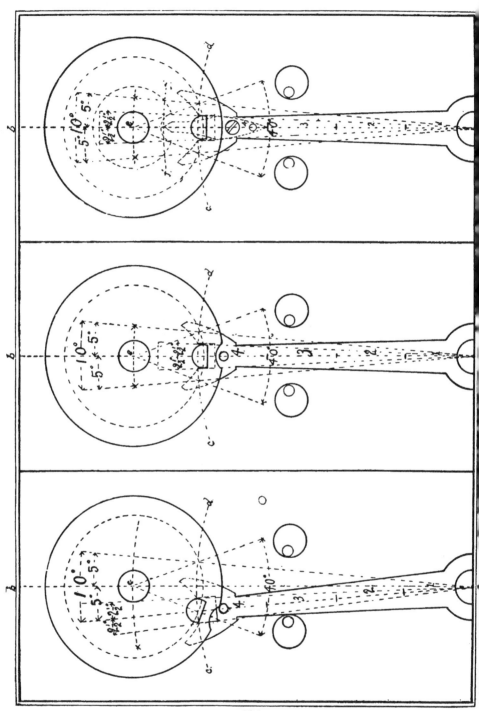

be the same, the roller being larger in diameter when the lever was longer, so in making our drawing, we may use for the length of the lever .6 or .7 of the diameter of the escape wheel.

The size of the roller and the length of the lever vary, a large roller and a short lever giving the balance impulse during a less number of degrees than a long lever and a small roller. The radius of the roller is made from ⅓ to ⅕ the length of the lever. In speaking of the length of the lever, we mean the distance from the center of the pallet arbor to the mouth of the notch the jewel pin enters. The horns of the lever projecting beyond this point, is a part of the safety action.

The lever and pallets in many cases are made of one piece; if not, they are securely fastened to each other, so they must move together. We found in making our drawing of the escape wheel and pallets, that the pallets had an angular movement of 10°; our lever must have the same angular motion, as both move together. In making our drawing, we will lay off 5° at the right and 5° at the left of the line a b (Fig. 10). These two lines will represent the position of the center of the lever when it has moved the pallets far enough to allow a tooth to pass off of the R. stone and also from the L. stone, a total movement of 10°, of which 1½° is the lock and 8½° the impulse or lift, as shown in the last drawing.

We next draw the arc c d which gives the length of the lever being .5, .6 or .7 of the diameter of the escape wheel; the width of the notch for the jewel pin is found by laying off 2½° at the right and 2½° at the left of the line a b, making the notch 5° wide. The sides of the notch are drawn parallel to the line a b. The depth is determined by the distance the jewel pin enters, it only being necessary that it should not touch the bottom of the notch.

To determine the size of the roller, we divide the length of our lever into 3, 4 or 5 parts (in this case 4 parts are taken); one of these parts gives us the radius of the roller, or the distance from the center of the jewel pin to the center of the roller. To locate the center of the roller, set the dividers one-fourth the length of the lever or one of the four equal

135

spaces, then place one of the points where the 5° line crosses the circle c d and draw an arc of a circle crossing the center line a b. Where this arc crosses the line a b, the center of our roller will be located, shown at e, in the drawing. Now place the dividers at this point, and draw a complete circle, which gives us the path of the *center* of the jewel pin and also cuts the arc c d at the points where the 5° lines cross it. The point where this circle crosses the line a b is the center of our jewel pin. The size of this should be as large as it can be and act freely in the notch of the lever. If the jewel pin is small, much power will be lost, so we may draw in a circle that nearly fills the notch. Some allow ½ of a degree for shake; this, I think, is rather excessive, a less amount giving better satisfaction. Jewel pins are made of various forms, the most common being the round ones with ⅓ of the front face flattened or ground off. One of this form enters the notch in the lever better and is more satisfactory than either the round or the oval ones, so common in English and Swiss watches. In some of the better grades of watches, the triangular ones are used and give very good satisfaction, although they are not as strong as the round ones with ⅓ of the front face ground away. The guard pin should be placed as near the bottom of the notch in the fork as possible, and allow strength of metal to hold it. When this has been drawn in, we may determine the full diameter of the roller, the farther from the notch the guard pin is located, the larger the diameter of the roller. It would be rather difficult to tell just how large to make the roller when the lever is at rest or in the center, but when it is at either side, then the edge of the roller would just touch the guard pin, so we may draw a circle showing the guard pin in the position it would occupy on either 5° line (as shown in the drawing at the left) and then draw a circle just touching it, which will give the full diameter of the roller. The crescent may now be drawn in and should be deep enough to allow the guard pin to pass through with a little clearance at the bottom, the width of the crescent may be taken from the points where the 2½° lines cross the circumference of the roller.

The inner parts of the horns of the fork should be drawn so they will be parallel to the circle representing the path of

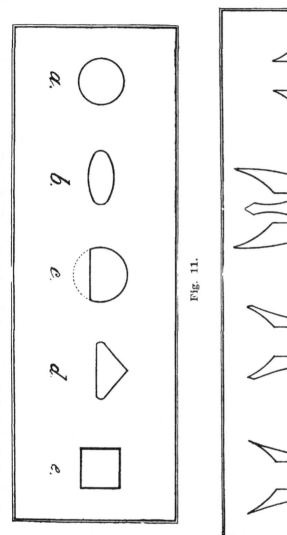

Fig. 11.

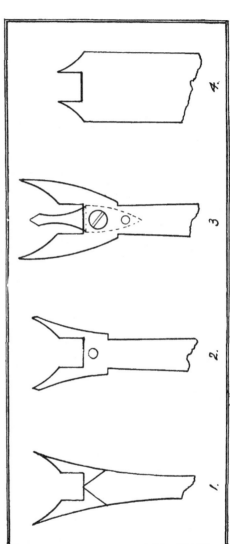

Fig. 12.

the center of the jewel pin when the lever is in the position shown in the drawing at the left, the curve of the horn which the jewel pin would pass in leaving the fork, would be drawn from the center of the roller. The other horn would be drawn from the same point when the lever was resting against the banking pin on the right. When the lever is at rest, its center being on the line a b, then in drawing the horn on the right side, we would place our dividers not at the point e, but 5° at the left of it, shown by the mark x, and the left horn would be drawn from a similar point shown on the 5° line at the right. The other parts of the lever may be made to suit one's fancy, some are very plain and others quite elaborate.

The drawing shows the banking pins in the position they are in when a watch is *banked to the drop*, in other words they allow the lever to move just far enough to let the teeth of the escape wheel to pass off from the impulse faces of the pallet stones, the side of the pins would be located one-half of the width of the lever from the 5° line as the center of the lever would be on that line after moving 5° from the center.

The lifting angle of the roller in this case is 40°, if the radius of the roller was one-third the length of the lever it would be 30°, etc.

In many of our modern watches of the better grades we find escapements with a *double roller*, the escape wheel and pallets are the same as in other escapements, the difference being in the *safety action*. In the fork and roller we have just drawn the guard pin comes in contact with the outer circumference of the roller, the farther this circumference is from the center the greater the amount of friction, in the double roller the guard-point projects beyond the end of the lever and is also below it; by this arrangement we can have a much smaller roller for the safety action, it being only one-half or three-fifths of the diameter of the circle representing the path of the center of jewel pin; on account of this very small diameter the friction of the guard-point is greatly reduced and the crescent being much deeper, we are able to get a much better safety action. The drawing at the right shows the double roller. It differs from the others, only in

138

the addition of the small roller and the guard-point. The size and shape of the large roller is of no importance, as its only purpose is to carry the jewel pin. They are made of many forms, some makers even setting the jewel pin in the arm of the balance the proper distance from the center. In our drawing, the small roller is three-fifths of the diameter of the circle showing the path of the jewel pin, the guard-point would just touch the small roller when the lever was either 5° at the right or left; an arc of a circle drawn through these points f f, crossing the line a b locates the end when the lever is at rest. The crescent must be deep enough to allow the guard-point to pass through with some clearance. The center of the jewel pin and the center of the crescent should always be on a straight line from the center of the roller.

In Fig. 11 is shown five forms of jewel pins that have been used; a was used in English watches, but on account of its being round, it would not enter the notch with the same accuracy as those in use at the present time. The Swiss people used those of the oval form shown at b, this was a step in the right direction, it being more flat, entered the fork better and had a safer action than the round one. At c is shown the one now in general use, it being a round one with one-third of its diameter removed from the front. This is the kind used in nearly all American watches and also in our drawings, the front face being flat, the jewel pin enters the notch deeper than the other forms, and consequently has a much better action. The trianguler one d, is found in some very fine Swiss and a few American watches; its action is similar to the last mentioned, being perhaps a trifle superior, the form shown at e is not in general use today, some of the American watches used them for a time, but for various reasons they did not give general satisfaction, and soon went out of use.

Fig. 12 shows four different forms of forks, and also their safety action. 1 is an ordinary Swiss lever with a dart formed by filing the front of the lever as shown, this dart acts in the same manner as an ordinary guard pin coming in contact with the edge of the roller. 2 is the common lever with the brass guard pin inserted. 3 is the guard-point used in a double roller, projecting beyond the notch, it is a piece fastened on the under side of the lever by a screw as shown. 4 is a lever without any guard pin, and is found in English watches that have a crank roller. In this case, the curved portion of the fork comes in contact with the outside of the roller, the jewel pin projecting beyond the circumference of this part of the roller.

We have learned the principles on which the escapements are made, and now we should be in a condition to begin a systematic study of the escapement, and learn its defects and how to correct them. When these drawings are understood, it will be surprising how much easier escapement problems will appear, as most people are working in the dark when they do not know why a thing is done. By the aid of many photographs showing the action of escapements step by step, also showing defects in construction or those caused by accident, it is hoped that my readers may be able to overcome any difficulties now existing, and future work may be a greater pleasure.

We have studied the theoretical principles of the lever escapement pretty thoroughly in making the drawings, and should now be in a condition to do some good *practical* work. It is impossible to do good work unless the theory is understood, yet theory is of little value without practical application; we cannot tell just how many degrees action the various parts of the escapement may have in practice, yet if we have formed a good image of the drawings in our mind, it will help us to quickly judge if the parts are in their proper relations to one another.

It will be best to consider the pallets and the escape wheel first, leaving the fork and roller action until later. Some people in setting the escapement, get the fork and roller action right first and then set the pallet stones to fit; this I think a very bad practice, as the roller may be too large or too small, and the guard pin as used in the American watches is very liable to be bent. Any of these defects would make it difficult to get the pallet stones properly set.

When the escapement was drawn, we began with the escape wheel and pallets. We allowed 1½° for lock, 4° for the impulse on the pallet stone and 4½° for the impulse of the tooth, making a total amount of 10° If the roller should be too large, then the lever would have more than 10° movement, and the result would be the pallets would have the same angular movement, and as we can not change either the impulse face of the tooth or that of the pallet (they being previously made), the extra amount of motion would increase the amount of lock, a very bad defect, as the extra force used in unlocking would be taken from the impulse.

The locking in all cases should be as little as we can

Fig. 13.

give it, and have it safe and sure. We allow 1½°, but if we can make our work accurate enough that only 1° is necessary, so much the better, but we must allow some for the side shake of the pivots in the jewels and we often notice a slight irregularity in the teeth of the escape wheel. The wheels when made are supposed to be true. Many a wheel that is true in the round, has been poorly fitted to the pinion, often decentered; we can not cement up an escape wheel and true from the teeth like we would a train wheel, the teeth being so delicate they would be liable to damage, so we must resort to a different method as follows: (Illustrated in Fig. 13.) Take a cement brass larger than the escape wheel and cut a step or recess nearly as deep as the thickness of the wheel and the diameter just large enough that the wheel may enter and have no side shake, the wheel may now be cemented in with a small amount of lathe wax or shellac, heating the cement brass hot enough to melt the cement when touched to it, before placing the wheel in position. It will be readily seen that the step being turned in the lathe, must be true, and the teeth of the wheel fitting this step closely, the outside edge of the teeth must be true, and now if we cut out the center with a small jewel graver, the hole must be true with the outside of the wheel, by bushing the wheel if too large, and cutting the hole in the bushing true, our wheel will be as perfect as it is possible to make it. After the bushing has been placed in the wheel, we can replace it in the cement chuck and cut the hole the proper size to fit the escape pinion with a *very small* jewel graver. Do not attempt to drill or broach an escape wheel and have it true. It should always be cut out with a small graver. I have often made small cutters out of needles for such work, grinding them the same shape as the jewel gravers shown in the previous article on jeweling.

A wheel out of true or decentered causes so many defects in the escapement that I urge the correction of such defects before any attempt is made to set the pallet stones. A wheel out of center would have heavy locking on one side and light on the other, or it might be correct on one side and have none on the opposite side. In any case, it would not be possible to get equal locking with all of the teeth. Another effect from the same cause would be a slight difference in the drop caused by the unequal lock of the different teeth.

One of the first things we should learn is how to select a pair of pallets to fit an escape wheel, or an escape wheel to fit a pair of pallets. This is not nearly so difficult as many imagine, and here our knowledge gained by making the drawings will be very valuable. We notice the tooth of the escape wheel is locking on the "R" stone, the fourth tooth (counting the one locking, as the first), the let off point is a very small distance from the let off point of the pallet; this amount is the drop, and is shown in the photograph (Fig. 14). When a tooth is locking on the "L" stone,

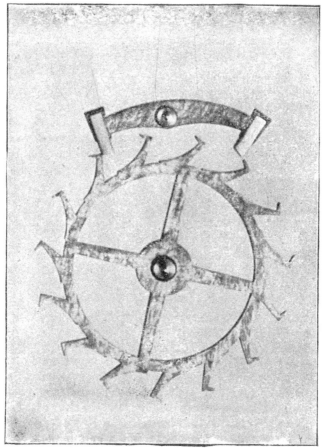

Fig. 14.

then the space between the let off point of the "R" stone and the back of the tooth that has just escaped from it should be exactly the same as the space between the "L" stone and the tooth that has escaped from it. These are the important points to observe in selecting either a wheel or the pallets. It will be noticed that in the first case the pallets fit *between*

143

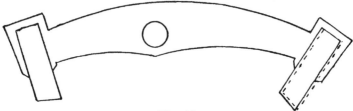

Fig. 15.

four teeth with a small amount of shake; in the second case, they fit *over* *three* teeth with the *same* amount of shake. When these conditions prevail, we may be quite sure the wheel and pallets will fit each other. They may now be tried in the depthing tool, having set it the correct distance between the jewels of the escape pinion and the pallot arbor.

When the pallets are too large for the wheel or the wheel is too small for the pallets, then the shake between four teeth will be very small, or perhaps the pallets are so large for the wheel that they will not enter the space between four teeth, and when the pallets are placed over three teeth, the shake will be excessive. It will be readily seen that we can tell almost at a glance by placing the wheel and pallets in position shown, whether one will fit the other or not. We may find many cases where the shake is unequal, and often do where the pallets may properly fit the wheel, the error being caused by the pallet stones being incorrectly set; for instance, if the stones are too close together the effect would be the same as pallets that were too small. The defects would be corrected by moving the stones farther apart, or by moving either *one-half* as much as we desire to increase the space. Our reason for moving only one-half of the amount is shown in Fig. 15. Suppose I move the "L" stone 1-10 of a millimeter to the right, as shown by the dotted lines; by doing so, we have increased the space over three teeth that amount; we have also decreased the space between four teeth the same amount, so if our inside space has been increased 1-10 of a millimeter and our outside space has been decreased 1-10 of a millimeter, we have a total difference of 2-10 of a millimeter. This accounts for what appears to be a large difference in action for only a very small difference in the movement of the pallet stones.

144

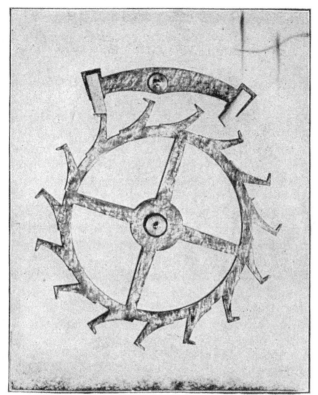

Fig. 16.

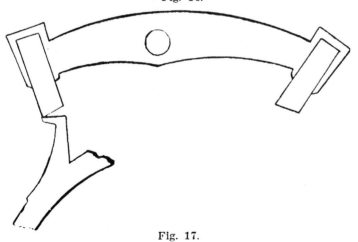

Fig. 17.

MODERN METHODS IN HOROLOGY.

We often find pallet stones set in the pallets with the impulse face of one or both stones in the wrong direction. There is no excuse for anything of this kind if one gives any thought to the construction of the escapement. The impulse face of the tooth and pallet should be nearly parallel where they come together, as shown in Fig. 16, and never as shown in Fig. 17, which shows each stone reversed. Here our impulse would be nearly all lost, only a portion of that on the tooth having effect. It is as necessary sometimes to show a defect in order to impress the correct form upon one's mind, as to show the correct form itself.

For a practical illustration, let us suppose we have a watch that needs new pallet stones, the old ones being either lost or broken. We order a pair for the movement, stating the size, grade and make. When they arrive, we will notice one of them has less angle than the other, the one with the least angle is the "R" stone and the one with the greater angle is the "L" stone. By examining any of the drawings or the photographs, one can see at a glance which way the stones should be set. The shortest side of the stone is the locking face; a tooth of the wheel should always drop upon this face after a tooth has passed from the impulse face of the other stone.

We can hardly expect to get the pallet stones set exactly correct the first attempt, we can only get them in their proper place, so that when we place them in their position in the watch, we can quickly tell how to move them to correct any faults. There are two things we must keep in mind: First, we want a certain amount of locking on each stone, and, second, we must have our watch in line; by that we mean the lever should pass an equal distance on either side of the line of centers between the pallet arbor and the balance staff. We test for the lock first by banking the watch to the drop. When a watch is banked to the drop, the lever and pallets can move just far enough to allow all of the teeth to escape and no more. The quickest way to bank a watch to the drop, is to turn one banking-pin in toward the center, and the other away from the center. The one near the center will not allow the lever to move far enough to let the teeth of the wheel escape, while the other one will have no effect

146

on it at all, as it is too far away. The balance wheel
should not be in place while banking the escapement. While
there is some power on the train, move the lever until it
rests against the banking-pin near the center. It will not
allow the tooth to escape; now with a small screw-driver
move the banking-pin very slowly away from the center,
watching the wheel carefully until we see it drop. Then
move the lever back and forth until the wheel makes a complete *revolution*, as some of the teeth, which is often the case,
are slightly longer than the others. When they all escape,
we will turn the other banking-pin toward the center, and

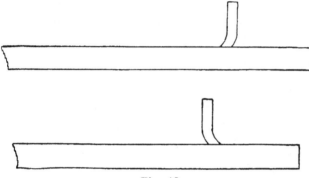

Fig. 18.

move the lever against it, open this pin until the wheel
escapes the same as before; when all of the teeth escape, our
watch is banked to the drop. Now, we can examine the
amount of lock on the pallet stones, it should be about one-
sixth of the width of the stone, or as little as possible, and
be safe. In some cases the stones will be so far back that
there will be no lock at all, the teeth dropping upon the impulse faces. The remedy would be to set *one* or *both* of the
stones farther out. If the locking is greater than the amount
required, we can correct it by setting one or both of the
stones farther back, banking the watch to the drop, of course,
every time the pallet stones are changed, it will be noticed
that in changing the amount of lock, it may be done by
moving one stone or by moving both of them. We must
know when to move one, and also when it is necessary to

147

move both of them. As soon as we get our watch banked to the drop and examine the locking, we should place the balance wheel with staff and roller (taking off the hair spring), in position, and examine the shake between the guard pin in the lever and the edge of the roller. We learned in making the drawing that there should be no shake between them when banked to the drop. If the locking on the pallet stones is the right amount, and there is no shake between the guard pin and the roller, our escapement is correct, or if there is shake between the guard pin and roller, and it is *equal*, our pallet stones are set correctly, but the roller is either too small or the guard pin is too far back. The safety action may be corrected by bending the guard pin forward, giving it a double bend (Fig. 18), so that the part coming in contact with the edge of the roller is always perpendicular to the lever. When this is not done, the difference in the end shakes will often cause the safety action to be faulty. If the guard pin should be tight on both sides, we would correct it by bending the pin back in the same manner as just explained, as shown in same illustration below.

We do not often set the pallet stones, and get the proper lock the first time. To do this, and also have it in line, is almost impossible. Suppose now after setting the stones and banking to the drop we find the locking too heavy and, on trying the shake on the roller, find one side just free and the other having considerable shake. We know that by setting back either stone will reduce the locking on both pallets as much as that one is set back. We also know the lever will be moved a less distance, so by setting back the proper stone we can correct the locking and bring the watch in line by one move.

Again we will suppose after setting the stones and banking that the locking is very light and the shake appears to be correct on one side, but the lever will not move far enough to allow the jewel pin to pass out of the fork on the other side. We know that by setting out the pallet stones it will give the fork a greater angular movement. We know that by setting out one of the pallet stones, both will have more lock; by setting out the right stone, we can correct our locking, also correct the fork and roller action, as we must open our banking to allow the wheel to escape, the lever will move farther, and this will let the jewel pin pass out of the fork, correcting our trouble, and putting the watch in line.

What was said about selecting a pair of pallets to fit an escape wheel applies equally well to the drop. A tooth when

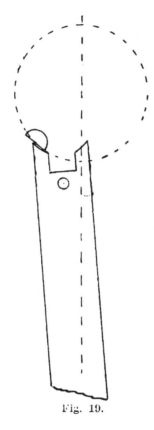

Fig. 19.

leaving the "R" stone should drop the same distance one does that leaves the "L" stone, otherwise we have unequal drop, and the watch would sound as though it were out of beat. If the pallet stones are too far apart, the *inside* drop will be excessive, and if they are too close together, the *outside* drop will be the greatest. They should be equal; the remedy is plain.

149

There are two ways of putting a watch in line; one has been shown, that of moving stones; another, which is possible in many watches, is accomplished by moving the lever on the pallets after setting the stones correctly. Some levers are held in place by screws, others have steady pins and some are made of the same piece as the pallets. These cannot be moved, and in some cases are bent in order to bring the watch in line.

In any case where the watch is banked to the drop, and the lever does not allow the jewel pin to pass out on one or both sides, never try to correct the difficulty by opening the banking pins, as this will not correct the trouble. A watch, to give good satisfaction, must allow the jewel pin to enter the fork and pass out freely when banked to the drop, and any escapement that will not do this should be made to do so. Sometimes it is the fault of the jewel pin, it being set out too far. It may be and often is the end of the fork, that is too long, the jewel pin striking the corner as it enters (Fig. 19). There should be a small amount of space between the outside face of the jewel pin and the horns of the fork, as may be seen in the drawing of the fork and roller in the last article. Many of the new watches as they leave the factory will not allow the jewel pin to leave the fork or enter it when banked to drop. This trouble may be corrected by grinding out the fork with a round iron wire and oilstone powder, afterwards polishing in the same manner with diamantine and oil or another wire of brass, as shown in Fig. 20.

The jewel pin should fit the notch in the fork as closely as possible and be free. This is very important, as much power may be lost at this point, and the setting of a jewel pin is a much more important item than many think. Let us see the duties it must perform: First, it enters the notch in the fork, moving the lever until the pallet stone unlocks, then the force of the train acting upon the escape wheel gives the pallets their impulse by the combined action of the impulse faces of the tooth and pallet stone (passing over each other). The jewel pin is no longer acting against the lever, but the lever is now giving impulse to the balance through the jewel pin. If the pin is too small to fit the notch nicely, then the lever will move some distance without giving impulse; many watches fail to have good motion on account of this defect. Many workmen select a jewel pin to fit the hole in the roller instead of fitting it to the fork, which should be done. In case the jewel pin is too large to enter the hole in roller, the hole may be enlarged by using a piece of binding

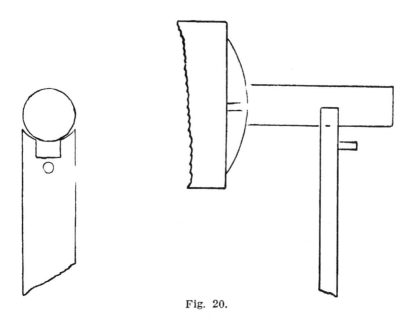

Fig. 20.

wire in the saw frame, and grinding it out with oilstone powder and oil. It is important that the jewel pin should be set square and perpendicular to the roller table; if it fits the fork as it should, the lever will give the balance im- pulse *immediately* after the tooth of the escape wheel unlocks from the pallet stone.

We have learned to draw the lever escapement properly, select our escape wheel or a pair of pallets, and should have a very good idea of the principles involved. There are, however, many little hints that will help in repairing that may be added at this time. Often in banking a watch we find the banking pins are too loose and have not friction enough to keep them from turning in the plate. When in that condition our fork and roller action would be affected,

151

as the pallets would have too much run caused by the banking pins moving outward by the constant striking by the
side of the lever. We often find cases where the threads
have been flattened with a hammer to tighten them, this
should not be done as the threads are very fine, and when
forced into the plate would destroy the threads in it. We
can correct a fault of this kind very easily and without
damage of any kind in the following manner: Make a
wedge-shaped punch as shown in Fig. 21, remove the banking
pin from the plate and place the pin in a hole of our bench
block, allowing the head to rest on the block as shown, then

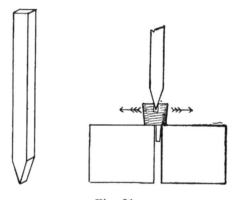

Fig. 21.

with the punch in the slot of the screw, strike it lightly with
a hammer and the head will be spread just enough to increase
the diameter of the screw, but has not disfigured the threads
in any way. We often find banking screws that are so tight
it is almost impossible to move them. Such should be taken
out of the plate and a little bees' wax put on the threads,
which will prevent the two pieces of metal sticking together. ·
Great care should always be used to prevent the pins being
bent, as a bent banking pin often prevents a watch from running in different positions. I have known several cases
where this alone would cause the watch to stop occasionally.
In many of the modern watches we find a fork made
of bronze or a composition metal that is quite soft; in these,

where the jewel pin constantly strikes in the notch, it is rapidly worn so that in a short time it greatly affects the motion of the watch. We also notice the same thing in some steel forks. When this occurs, the notch should be ground out and repolished, and a larger jewel pin placed in the roller or the lever notch may be closed and ground out to fit the jewel pin.

The teeth of the escape wheel constantly dropping upon the locking faces of the pallet stones, often wear a little notch in the stone. This makes the unlocking more difficult, and consequently lessens the motion of the balance; a defect of this kind can be easily remedied if we possess a diamond

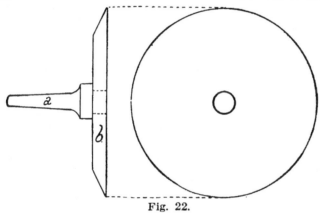

Fig. 22.

lap and understand its use. As these laps are quite expensive and many workmen would not feel that they could afford such a luxury, a method will be shown that any one may be able to make one at very small cost. The best laps are made either of soft steel or copper, the latter being mostly used. Fig. 22 shows a side and front view of such a lap, the steel taper —a— is turned to fit our taper chuck nicely, after which the shoulder is turned to fit the copper —b— which is soldered to shoulder of the taper and turned perfectly flat and round, it being a good idea to bevel the back as shown. It will be much better if one has a slide rest to do the turning, but it can be done with a hand graver nearly as well. The front face should be ground perfectly flat on a piece of ground glass with either fine emery and water or tripoli and

153

oil. When this surface is flat and smooth, the lap should be charged with No. 4 diamond powder. This can be bought for about $2.50 a carat and ⅛ of a carat will be enough to charge a small lap. I have made some very nice ones out of Canadian pennies. They are large enough for ordinary work, but any piece of copper about that size will do as well.

The best method of charging the lap is to mix the diamond powder with oil and force it into the soft copper or steel with a very hard steel roller. The softer metal retains the particles of diamond powder. If one has no steel roller, some other piece of very hard steel may be used, being careful not to leave the surface rough; a knife blade may be used for this purpose. We should also have another lap of similar shape made of box-wood for polishing. This we do not charge, but use with putty powder (oxide of tin) and water for polishing.

After these laps have been in use a short time, no one would be without them for many times their cost. In using the diamond lap, the surface should always be well oiled or moistened with water. In no case should it be used dry; any rough or chipped pallet jewel may be ground smooth with the diamond lap and afterwards polished with the box-wood and putty powder, making them as good as new. Perhaps no greater use will be found for this lap than in grinding off the end of jewel pins. We often see them set projecting through the top of the roller, and in many cases they are so long they drag on the plate below. The jewel pin should be set even with the surface above, and if too long, may be ground off until it is just long enough to reach through the lever; this can be quickly accomplished with the laps described.

A fault that exists in many escapements, and is often overlooked, is a roller that the edge is poorly polished. It may have been well polished when made, but some careless workman may have marred it in removing the roller from the staff. If this edge is not highly polished and the guard pin comes in contact with it while the watch is running, it is liable to stop very quickly; often in setting a jewel pin, a small particle of shellac may remain on the polished surface of the roller or in the crescent, and this will have a similar effect to the rough roller. It is a very good plan to take a piece of peg-wood and run it around the roller. If there is any roughness, it can be quickly detected; when the guard pin and roller act together properly, the moment the guard pin comes in contact with the edge of the roller, it will be repelled, and the draw given the pallet stones will immediately bring the lever against the banking pin, leaving the balance and roller free to make its vibration.

We often find a roller badly out of true, usually caused by the hole being closed to make it fit a staff that is too small. This is a very serious defect, and the watch is liable to over-bank. A new roller is the best remedy, but we do not always have one of the correct size on hand, and often there is not time to order one from the material house. If the outside of the one we have is in good condition, we can make it answer the purpose as well as a new one with but little labor, the method being very much like the one used in bushing the escape wheel. Take a cement brass or a piece of wire larger than the outside diameter of the roller, cut a recess in which the roller will fit without side shake and cement it in. Now, cut out the hole in the center until it is perfectly true, it will, of course, be too large for the staff. We may now turn up a bushing of nickel or brass, the former making the neater job, with a hole that will fit the staff. This method of truing from the outside by turning a recess in a brass cement chuck is one that will be of use to us very often for various kinds of work, it may be quickly done, and no other method can be more accurate.

Our study of the lever escapement would not be complete without some knowledge of making the escape wheels and the pallets. The former will be taken up later under wheel cutting, and the latter will now be explained. We could not lay out an escapement the actual size on a plate and get it accurate, but we can make a large drawing on paper either ten or twenty times the size we wish our finished work. When this is done, it may be transferred to the plate from which we are to make our pallets.

We first make the drawing of the escape wheel and pallets the same as the one before explained, only making the parts ten or twenty times the size. The only additions are

155

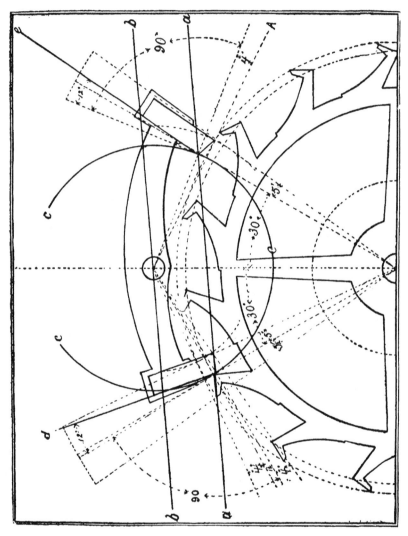

Fig. 23.

the two parallel lines shown (Fig. 23) a a and b b. First, draw the line a a through the locking corner of each pallet stone, the line b b will be drawn parallel to a a, passing through the center of the pallet arbor. The circle c c c also passes through the locking corner of both pallets. When this

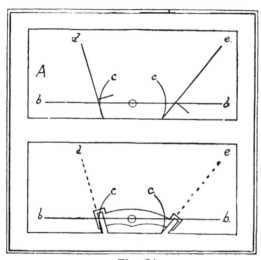

Fig. 24.

drawing is made, we will proceed as follows to make a pair of pallets of actual size: A (Fig. 24) is a piece of sheet steel a trifle thicker than required for the finished pallets. The surface should be quite smooth that the lines may be distinctly seen. The working edge at the bottom in this case should be filed straight and represents the line a a in the drawing. We will suppose the drawing has been made twenty times actual size. We now draw a line parallel to the bottom edge one-twentieth of the distance between the lines a a and b b in the drawing, shown at b b in Fig. 24; at some point on this line near the center make a mark with a very sharp center punch, which will be the center of the pallets and pallet arbor. From this point, draw a circle one-twentieth of the diameter of c c. In Fig. 23 where this circle crosses the line a a, the locking corner of the pallets will be located.

We are now ready to locate the locking faces of our pallets: First, take a very thin piece of sheet metal and cut

157

and file it carefully to the shape shown in Fig. 25. This may be done by placing the bottom a a on the line a a in the drawing, and filing the side a d so it will be the same angle as the locking face of the R. stone. The side a e will be made in a similar manner corresponding to the locking face of the L. stone, this piece being made quite large will determine

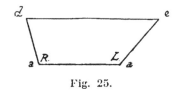

Fig. 25.

the angles very accurately, and may be used on any size pallets, as these angles are the same in either a large or small pair of pallets.

To mark our locking faces, place the side a a so it is just even with the bottom of the steel and the corner "R" at the point where the circle C crosses the edge of the plate, and draw the line d, which will be the locking face of the "R" stone. Now, place the corner of this metal piece marked L at the point where the same circle crosses the edge on the right and draw the line e, which gives the locking face of the "L" stone. The length of each stone should now be shown by drawing a line at right angles to each locking face as shown, when this has been laid out on the surface, the remaining work may be quickly accomplished. We select a saw, either circular or an ordinary one, which is one-twentieth the width of the pallet stones in our drawing; with this saw cut the slot for the pallet stone at the right of the lines d and e, being careful to keep the edge of the cut on these lines, the depth being determined by the lines at right angles.

The slots for the two pallet stones are the most important things in making the pallets, as the form depends very greatly upon the makers' individual taste.

Should any one desire to make the pallet stones also, they may be ground to fit the slots, taking the angles and sizes from the large drawing.

This is as simple and correct a method of making the pallets as any that has ever come to my notice.

158

THE MAINSPRING.

The motive power of a watch depends upon the elasticity of a long, thin ribbon of steel, called the main spring. Its action seems to be less understood than any other part of our pocket time-pieces; it breaks without any apparent cause, often soon after being placed in the barrel; again, it will last for many years and perform its duty as well as when new.

The spring of an ordinary 18-size movement is about twenty-two inches long and one-eighth inch (about 3 mm.) wide. The thickness varies with the quality or grade of movement, the weakest or thinnest being used for the highest grades, as they require less power to run them. A wide, thin spring is less liable to break than a narrow, thick one, so in constructing a watch, the barrel is made as large in diameter as possible, and as thick as the space between the plates will allow. In many Swiss watches the barrel arbor has its support all on one end; this permits the barrel to pass completely through the other plate, allowing a very wide spring to be used, and when well made, gives good satisfaction, but all parts must fit closely to prevent any side shake to the arbor.

One of the most difficult tasks of a watchmaker is to convince a customer that a spring will break without any apparent cause, even while the watch is not being carried. They, of course, naturally ask us what makes them break; then, when we are unable to tell the exact cause and why it did break, they are inclined to distrust us. The real cause *may* be known in the future; surely here is a good field for research and a fortune for the one who will discover the cause and invent something that will prevent the breakage. When we stop and think of the very small space in the barrel that this spring must occupy, and the still smaller space it occupies when fully wound up, that it is wound up and runs down day after day for years, we will wonder that it can stand the strain as long as it does without breaking.

159

Often a spring will break soon after being taken out and replaced while cleaning, even when it has been in use before for years. For this reason some good workmen do not remove the spring from the barrel in cleaning, but if the oil is thick and gummy, it should be taken out and thoroughly cleaned, as upon the good action of the spring depends much of the motion of the watch.

A spring should have its surface nicely polished, and should be well oiled, as in unwinding the coils slide upon each other, and should have as little friction as possible.

No attempt will be made to explain why a spring will break, yet there are many little things that are overlooked by

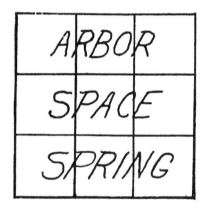

Fig. 1.

most workmen which may be brought to our notice, and might prevent breakage in some instances.

Comparatively few workmen understand how to fit a spring properly. Most of them gauge the old one and put in one the same as the one that was broken; in many cases this will be satisfactory, but suppose the old one is not the proper one; should we make an error because some other workman has done so? No. Let us understand our work so well that we may be able to correct any fault found and not be in error because others have been.

Theoretically, the arbor should occupy one-third of the *area* of the inside of the barrel. The spring should occupy one-third and the space should be the same. This may be better illustrated by Fig. 1. Here we have a square diagram divided into nine equal spaces; the upper three or one-third of the area represents the amount of space filled by the arbor. The next three represent the amount of *space*, while the bottome three gives us the amount of space occupied by the spring. This is easily understood when represented by a square diagram, but in actual practice where the barrel is round we have a different problem. We will not consider the size of the arbor in fitting the spring, as that is made to fit the watch and its size cannot be well changed. We

Fig. 2.

must endeavor to get our spring to fill one-half of the *area* of the space between the arbor and the inside of the rim. At the same time, we must have a certain number of coils in the spring.

A spring that is too long will prevent a watch from running its full time, as well as one that is too short. In other words, if the spring has too many coils and fills up the barrel

161

more than one-third of its area, it is impossible for the arbor to make the required number of turns in winding. The arbor should make four and a half or five turns in winding up the spring. The barrel will, of course, make the same number of

Fig. 3.

revolutions as the spring runs down. Suppose the barrel has 80 teeth and the center pinion has 12 leaves; then while the barrel makes one revolution, the center wheel will make

Fig. 4.

80 ÷ 12 or 6⅔ revolutions; the barrel makes five revolutions. As the center pinion carries the minute hand and makes one revolution every hour, a watch so constructed would run 33⅓ hours with five turns of the arbor in winding. In many

cases the arbor will make more than this number, and consequently the watch will run a longer time.

The photograph shown at Fig. 2 gives a very good idea of the correct proportions. Here we have ten and a half coils in the spring, and the area of the space and that of the spring are the same, as the outside coil of the spring, when fully wound up, occupies the position of the outer edge of the space when fully run down. To study a spring, an old barrel may be cut out as shown in the photograph; make a mark on the outside of the barrel at the inner coil when run down; now, wind up the spring, and if the arbor makes the required number of turns and the outside coil comes to the mark, the spring fits the barrel if it has about ten and a half or eleven coils before winding. The spring might fill the right amount of space and have too many coils. This would show at once it was too weak, and if there should not be coils enough and the amount of space filled is correct, then the spring would be too strong.

Fig. 3 shows a photograph of a spring found in a Swiss watch that would not run twenty-four hours. There are

Fig. 5.

about sixteen coils and the spring filled up too much of the space. After breaking off the outer end so that the spring had eleven coils, the watch ran easily thirty-three hours. The barrel hook in this case was also defective as can be seen by the illustration. It is so long that it projects through the outer coil far enough that the second coil rests against it. The hook should be just long enough to barely pass through the spring, and should be undercut, the hole in the end of the spring should not be straight through, but slanted so it will lock securely on the hook. The temper of the spring should be drawn at the outer end for a short distance. The width of the spring should be as great as it is possible to use, and allow clearance between the bottom and cap, 1-10 of a millimeter being sufficient in nearly all cases. A very quick way to test the old spring to see if it was the proper width is to break off a very short piece about 10 millimeters long and place it against the inside of the barrel. If the cap is not countersunk the spring should be as wide as the distance

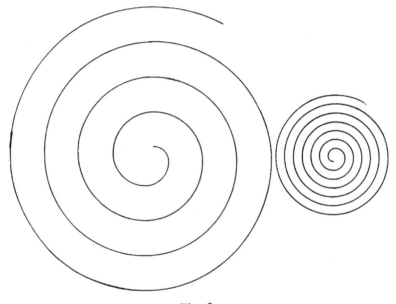

Fig. 6.

from the shoulder the cap rests against to the bottom, allowing a very small amount for freedom. If the head is countersunk, like many Swiss watches, we can use a spring as wide as the distance from the shoulder to the bottom, the recess in the cap being sufficient for the necessary clearance or freedom.

In placing the spring in the barrel, some good mainspring winder should be used, as when put in by hand they are often so badly bent that a good portion of the power is expended by the pressure against the top and bottom of the barrel, causing excessive friction, and consequently poor motion to the watch. Such a spring is shown in Fig. 4 as taken out of a movement that did not give good time. Could you blame it?

Another peculiar method of repairing a spring is illus-trated in Fig. 5. The spring had been broken and spliced as shown, and was in actual use for several years. I am not able to tell how good time the watch kept, however.

Some springs lose their elasticity after being in use a short time and become "set"; in other words, the coils fail

Fig. 7.

to open out properly when removed from the barrel. When one becomes set, it should be replaced by a new one. Fig. 6 shows a spring where the coils do not open properly, and also one where the coils open as they should.

We should use great care and not allow the fingers to touch a spring, as the perspiration will cause it to rust, and soon break. All mainsprings should be well oiled, clock oil being better than watch oil for this purpose, it being a trifle thicker, the coils constantly sliding upon each other require plenty of oil. A spring that is not well oiled is liable to break in many pieces. Fig. 7 shows such a spring. Why they break in so many pieces it is not easy to explain. It was my experience at one time to have seven springs break in the same watch in one-half of a day, and in each case the spring was in several pieces; they were all genuine American springs,

Fig. 8.

the best we could buy. There seemed to be no reason for their breaking. The steel may have been burned in hardening. Fig. 8 shows a photograph taken with the microscope, showing the end of a spring that broke into many pieces.

Fig. 9.

The grain is quite coarse and looks as though it was "burnt" in hardening. This would make it very brittle.

Fig. 9 shows a similar photograph, showing the end of a new spring of the best quality obtainable; the grain here is finer and less crystallized than in the other.

It is not difficult to find springs that properly fit American watches, as they are made the right length and the correct strength for each kind of movement, but we often see a spring intended for one make used in another and often causing serious trouble.

The T end of the spring should be carefully fitted, as it often projects through the top or bottom far enough to either touch the balance wheel or the center wheel. The T should be made the same length as the thickness of the barrel before the spring is wound in the barrel, as it is very poor practice to file it off afterward and disfigure the finished surface.

THE
COMPENSATING BALANCE
AND PENDULUM.

It is a well known law of physics that all metals expand in heat and contract in cold, but some to a greater extent than others. This fact made it very difficult to closely time the first watches made, as the high and low temperatures greatly affected their rate. This same knowledge of the effect of heat and cold on the different metals and its proper application have made it possible to time the modern watches so they have no perceptible variation in high or low temperatures.

It was generally known that a common clock would run fast in the Winter and slow in Summer; the pendulum rod was made of steel or iron and the cold of Winter would contract or shorten it, causing the pendulum to vibrate more rapidly; the heat of Summer would expand or lengthen the rod, when it would vibrate slower. It was impossible to get an even rate.

Many ingenious methods have been invented to overcome this difficulty. In nearly every case the unequal expansion and contraction of the different metals used is the principle involved.

One of the first attempts at correcting the temperature error was the use of the compensating regulator. This was made of brass and steel brazed together, one end being fastened to the regulator, the other end being free to move by the action of the heat or cold and also carried one of the regulator pins, one pin being solid and the other movable. When the pins were far apart the watch would run slower and when close together it would run faster. This action was automatic. When the watch had a tendency to gain on account of the temperature the regulator pins would separate, and when running slow the pins would be brought close together. In this way a closer rate than formerly was obtained. A balance cock with such a regulator is shown in Figure 1. The balance was a solid three arm one.

167

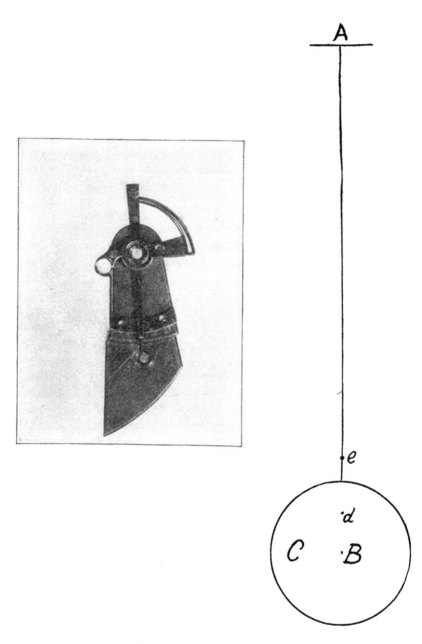

Fig. 1.   Fig. 2.

THE COMPENSATING BALANCE AND PENDULUM.

The principles involved in either a compensating balance or pendulum are nearly the same. It is a good idea to study them together. First, let us consider a simple pendulum (Fig. 2). It is suspended at A, the weight or pendulum bob is shown at C, the center of oscillation at B. Each particle of metal above the point B has a tendency to quicken the vibrations, while all particles below that point retard them. If a small piece of the metal could be removed from the point D and placed at E the vibrations would be faster, as the center of oscillation would be raised above B, consequently the pendulum would be shortened. This principle will be spoken of again in another form.

The gridiron pendulum is used in most Swiss regulators, although in many cases they are made for appearance and not for use, the same as many of their balances. When a pendulum of this kind is properly constructed it will give very good results. The drawing in Fig. 3 shows very clearly how the rods of steel and brass are arranged. There are five steel rods and four brass ones. The double lines represent the steel and the heavy solid ones the brass. The piece at the top, A, is connected by the two outer steel rods to the cross-piece B. Then the two brass rods extend from B to C. The next two steel rods extend from C to D. From D the next two brass rods extend to E and the center steel rod, which supports the weight, extends from E through the other pieces (freely) to the pendulum weight. The steel rods expand downward and the brass ones expand upward, and when properly adjusted the center of oscillation remains at the same point, consequently there will be no variation in the length of the pendulum at different temperatures.

The mercurial pendulum is used largely in the best regulators and makes a very showy as well as a very reliable one. Here we have two metals, but one of them a liquid. In most cases glass jars are used to contain the mercury. The rod that supports them is made of steel, which expands or lengthens in heat, and of course contracts in cold. The mercury in the jars expands upward, raising the center of oscillation, the rod lengthening lowers this same point. The smaller the glass jars are in diameter, the more would the mercury rise in heat; if too small, the pendulum would over-

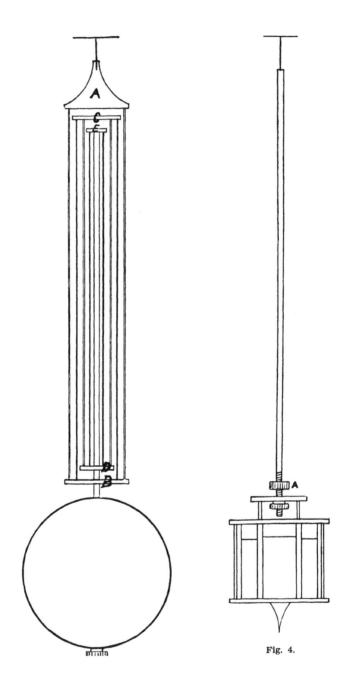

Fig. 4.

THE COMPENSATING BALANCE AND PENDULUM.

compensate. This pendulum is shown in Fig. 4. The nut at the end of the rod is used to raise and lower the mercury jars, or for the first timing, but when we obtain a very close rate to move this whole mass of weight, little enough would be hardly possible, so the extra nut shown at A is used. The principle alluded to at the beginning (Fig. 2) is now utilized. This nut as it is turned does not raise or lower the mercury or change the length of the pendulum, but raising the weight of this nut by screwing it up has a tendency to quicken the vibration or to lower it retards them, as by changing the position of this nut only, the center of oscillation will be changed.

I was asked some time ago to settle a dispute. A fine regulator had been placed in one of our public buildings by a local jeweler, but he was not able to regulate it closely. It had a losing rate, and he removed some of the mercury from the glass cells in the pendulum to make it run faster. The gentleman in charge of the building being pretty well posted with the sciences, told the jeweler he thought the clock would go slower when the mercury was taken out. I was asked to settle the matter, and explained that by removing the mercury the center of oscillation would be lowered, making the pendulum longer, and the clock would go slower. The mercury was replaced, and the clock was soon regulated.

Some clocks having a pendulum rod of wood well varnished to keep the moisture of the air from affecting it, keep remarkably good time without any compensating arrangement.

THE COMPENSATING BALANCE.

We have learned that the vibrations of a pendulum depend upon its length; with a balance we learn that they depend upon its diameter, the larger diameter corresponding to the longer pendulum and vice versa. So if we have a balance and move some of its weight farther from the center, it will go slower; or move a portion of its weight nearer the center, it will go faster.

The compensating balance is made of two metals, usually of steel and brass, having different expansive coefficient. The metal that is affected most is placed on the outside of the balance. The steel and brass are either brazed or fused

171

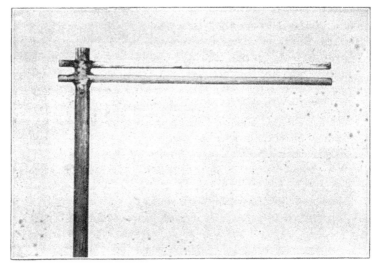

Fig. 5.

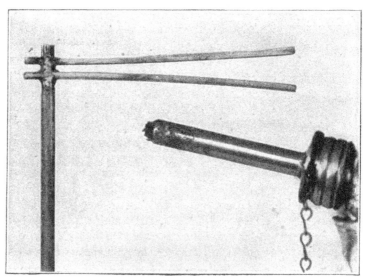

Fig. 6.

together so the two metals are practically one piece. To illustrate the action of the balance more clearly the article shown in Fig. 5 was made; the upright rod supports two straight pieces. Each of these pieces is composed of a strip of brass and steel soldered together with silver solder, the brass being on the inside in each case. At a normal temperature the pieces are parallel, but apply heat to them by the aid of an alcohol lamp, as shown in Fig. 6, the ends immediately begin to separate. The brass expands more than the steel, tries to pull the steel forward. The steel, on the contrary tries to hold the brass back, as it does not expand so much, so they form a compromise and the pieces are bent in a curved direction. Were it possible now to place

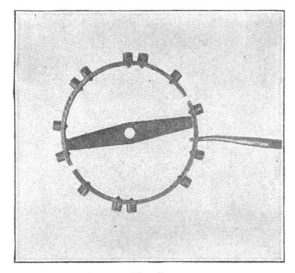

Fig. 7.

these same pieces in extreme cold, then the ends would come together, as the brass would contract more than the steel. This illustrates very plainly the action of the compensating balance, only we have a round rim instead of the straight piece just shown.

In order to show the actual effect of the different temperatures on a balance, the photographs in Figs. 7, 8, and 9 were made. One of the screws was removed and in its place a wire was used to support the balance while subjecting it to the different temperatures. Fig. 7 shows the cut on opposite side of the rim from the other two, being photographed from the under side. It also shows the rim in its normal position at a normal temperature. In Fig. 8 we see the balance and a portion of the alcohol lamp. The flame does not show, as it is nearly colorless. The rims are thrown toward the center. It will be noticed that the one nearest the flame of the lamp is bent toward the center the most; also that it is not perfectly clear, as it moved by the action of the heat a

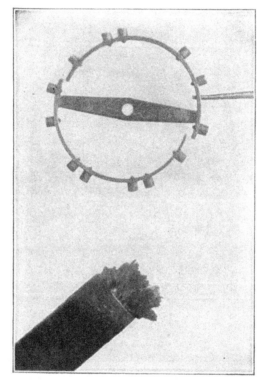

Fig. 8.

trifle while being photographed. In Fig. 9 we have the effect of extreme cold on the wheel. In this case the rim is thrown from the center.

Now let us see what effect the action of heat and cold will have on the time keeping qualities of the watch. In heat we notice the free end of the rim comes nearer the center, carrying with it the screws or weights. This acting the same as a shorter pendulum, would cause the balance to vibrate faster, and in the cold the rim would be thrown from the center, causing the balance to vibrate slower. When a watch runs fast in heat and slow in cold, it over compensates. In other words, the balance more than overcomes the effect of heat and cold on the hair-spring and other parts. If it should lose in heat and gain in cold, it would then under compensate. In either case the error may be corrected

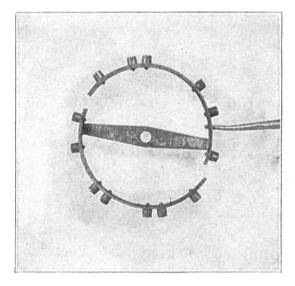

Fig. 9.

by changing the position of the screws. By inspecting the rim we find several holes that have no screws in them, and also in many balances the holes are closer together near the end of the rim. This allows the screws to be moved a very small amount.

In Fig. 10 we have a photograph of a modern balance wheel. The position of the screws would indicate that the tendency was to under compensate, as the weight is mostly near the end of the rim. If we should take the pair of screws marked 3 and place them between 4 and 5, the watch would run faster in heat and slower in cold than at present. There is one point often overlooked. The arms of the balance lengthen in heat and shorten in cold. By this action the whole rim would be thrown away from the center or brought nearer, causing a losing or a gaining rate. At the same time, while the arm is expanding and moving the rim from the center, the action of the heat on the two metals in the rim

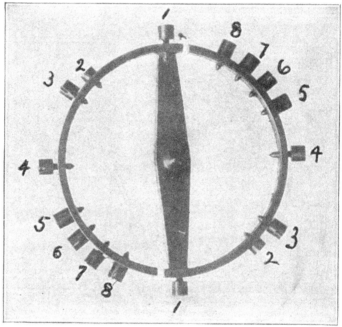

Fig. 10.

cause it to be brought nearer the center at the end, so there must be some point that is nearly stationary. This is called the neutral point and a screw placed at that point would have no effect on the temperature adjustment. This point will be very close to the small screws marked 2, and many watches are brought to time by placing screws at this point of the required weight. We should bear this in mind, and if at any time we find it necessary to change the weight of an adjusted watch, do it as near this point as possible, for we can easily see that altering the weight of any screw near the end of the rim would also alter the adjustment.

In the same illustration we see four screws—1, 1, 4, 4—that the heads do not come in contact with the rim. They are at the quarters and are sometimes called the quarter screws, but more often the timing screws. The threads fit very closely, and they are used for timing the watch. By unscrewing or turning them out the weight is moved farther from the center, which will cause the balance to vibrate slower. If they are screwed toward the center, the balance will vibrate faster. We should always move the two opposite ones the same amount, otherwise the wheel would be out of poise. We may move one pair or both pairs as we desire.

If the balance over compensates, we must move the screws toward the solid end of the rim. If it under compensates, the screws must be moved toward the free end of the rim. There must be some point between these two places where the rate will be the same in high and low temperature.

In order to obtain a good rate, it is important that the balance should be carefully poised. This is often overlooked. The wheel should be perfectly true before poising, for should it be trued afterward the wheel would be thrown out of poise again.

The balance should be in poise before any screws are put in, then each pair should be of the same weight. It is not necessary that all screws should be of the same size. To make this point more clear the drawing shown in Fig. 11 was made. At A we have a rod with a one pound weight on each end. One would just balance the other, it would be in poise.

At B we have two rods at right angles to each other. There is a one pound weight at each end of the horizontal rod and a ten pound weight at each end of the vertical rod. Each of these would be perfectly balanced. And at C we have three rods, with 1, 5 and 10 pounds. This one would also be perfectly balanced or poised, as the weights opposite each other are exactly the same. We may have some large and some small screws as long as each pair are of the same weight. In poising a balance that is known to have a close rate, it is best to add a small amount of weight to the light

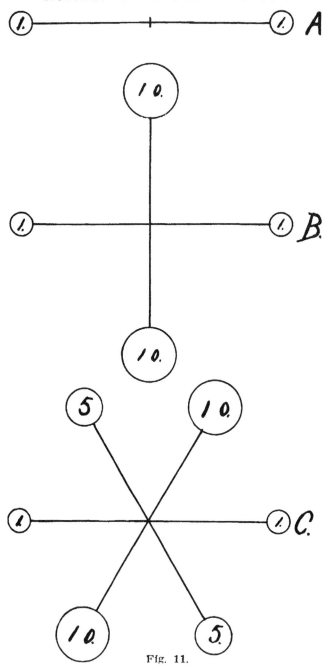

Fig. 11.

side and take away an equal amount from the heavy side. By so doing we will not alter the rate, which would be done if we should take it all off from one side. To reduce the weight on the heavy side we may undercut the head of the screw, as shown by the dotted lines in A Fig. 12, with a long pointed graver, or a still better way is to mill them out, as

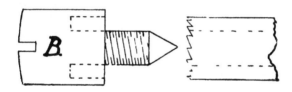

Fig. 12.

shown at B, same illustration. Either of these methods does not disfigure the screws in any way. If we find it necessary to add a small amount of weight, we may do so by placing a timing washer under the head of the screw, always selecting one of the same diameter as the head. Should the thinnest ones be too heavy, we can easily grind it down on our ground glass with oil and oil stone powder to any required weight. By any of these methods we do not alter the appearance of the balance in any way. I will show a method of poising that came to my notice a short time ago in Fig. 13. This is not

179

an example for any of my readers to follow, but shows how little respect some workmen have for either themselves or their work. The screw was filed on all sides in order to

Fig. 13.

make it lighter. There is no doubt about it doing the work, but oh! how it looks!

If the balance is nearly in poise and requires but a small amount of weight removed, we can cut the slot a trifle deeper with a very thin screwhead file.

A small pair of scales with grain weights will be found very useful when altering the weight of screws and poising a balance.

THE CYLINDER ESCAPEMENT.

As long as the cheap cylinder watches are sold by dealers, the watch-maker must be prepared to repair them, and in order to repair them, he must understand the principles of their construction.

I will not attempt to go into detail as much as with the "lever escapement," but will explain more particularly the points that will be helpful in repairing any that come our way.

The cylinder or horizontal escapement, as it is often called, is a dead beat escapement, i. e., it has no recoil in unlocking. A tooth of the escape wheel is acting against some part of the cylinder at all times except during the drop.

The cylinder is a very delicate part of the escapement; it is made of steel and must be very hard to stand the wear,

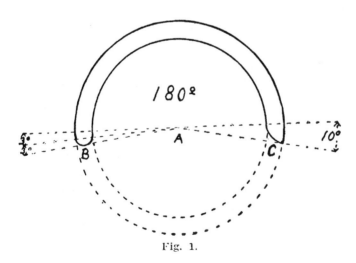

Fig. 1.

consequently it is very brittle and easily broken as the shell is only tempered to a straw color; it is on account of their hardness that they are so easily broken.

To better illustrate the shape of the acting part of the cylinder, reference is made to Fig. 1. A represents the cen-

181

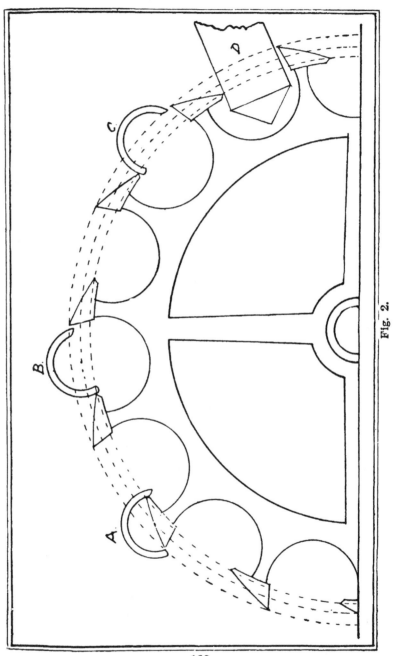

Fig. 2.

ter of the pivots on which it rotates. B the receiving lip and C the let off lip of the cylinder. The drawing is a cross section, and is a trifle more than half of a circle, as shown by the dotted line it embraces an angle of about 196°. The shell is very thin, and we find that this thickness varies greatly in cylinders of the same outside diameter, some having quite thick shells and others very thin ones; for this reason two cylinders having the same diameter will not always work in the same watch as we will see a little later.

Fig. 2 will give us a better idea of the size of the cylinder in regard to the teeth of the escape wheel. At A a tooth is shown inside of the cylinder, at B the cylinder is shown between two teeth of the wheel. It will be noticed that in each case they do not fit closely; while a tooth is inside of the cylinder, there should be some play, also while the cylinder is between two teeth, there should be play. When this play or shake over a tooth and between two teeth is the same, the cylinder is the correct size for that wheel; the point previously mentioned may be now understood. If the shell was very thick and the cylinder fitted between the teeth with the necessary freedom, and we would try the tooth inside of the cylinder, we would find it would not enter, or if it would, the outside and inside shake would be unequal, consequently we would have unequal drop.

A cylinder as we buy them is composed of four parts:

a. The cylinder, a hollow steel shell partly cut away in two places; the upper portion not being cut quite half way, thus forming the two lips which are rounded and polished. The lower portion is cut away about three-fourths of its diameter; this is to allow the rim of the wheel which supports the tooth to pass through, while the balance makes its vibration.

b. The top plug which fits in the cylinder tightly, fills the space from the top of the upper opening to the end of the cylinder and projects beyond far enough that the upper pivot may be turned upon it.

c. The lower plug which fills the space from the bottom opening to the lower end of the cylinder and projects below far enough to turn the lower pivot.

d. The brass collet which fits tightly upon the top end of the cylinder, and is turned to fit the balance wheel and the hairspring collet. Fig. 3 shows these parts clearly. As the top plug b and the brass collet d fit the cylinder closely, and are held in place by friction; the former may be driven out to

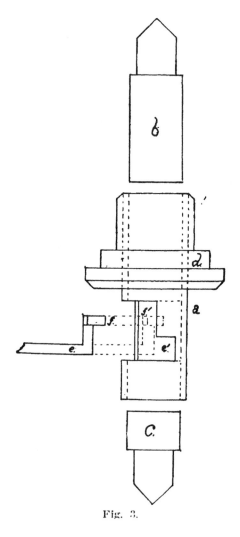

Fig. 3.

increase the length or to turn on a new pivot if the old one is broken, while the latter may be either driven up or down as the height of the balance requires.

184

THE CYLINDER ESCAPEMENT.

In the same illustration, the height of the escape wheel is shown with reference to the cylinder. The web or rim of the wheel e should be of such a height that it will pass through the center of the lower opening in the cylinder e', then the height of the tooth f will work nicely on the lips above f'.

In order to convey a still better idea of a complete cylinder, the photograph of the finished one shown in Fig.

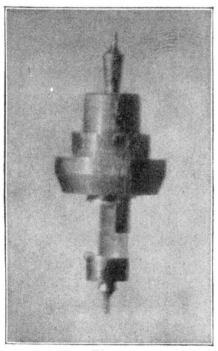

Fig. 4.

4 was made. This one is greatly enlarged, and gives a good idea of the general proportions. I also show a cross section of a cylinder through the lips. This is the part that is cut away nearly one-half; the receiving lip is rounded on both sides while the let off lip is only rounded on the inside. The reason for this is, the tooth of the wheel acts upon the outside of the cylinder near the receiving lip, and as it gives impulse, the action takes place upon both edges of the lip, while with the let off lip, the tooth gives impulse by acting

185

against the inside portion of the lip until the wheel drops.

In fitting a new cylinder to replace a broken or faulty one, we will proceed about as follows:

First, we must determine the size or diameter of the cylinder; if the old one worked properly, we may use one of the same diameter for trial. We should very carefully try

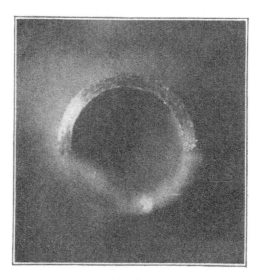

Fig. 5.

the shake between two teeth and over one tooth; if it is equal in both cases, the cylinder should do. It may be that the old cylinder is lost or not of a proper size, then we must proceed differently, or if we are obliged to send to a material house for the new one, we would turn up a small piece of brass wire, so it will fit between two teeth of the escape wheel with a little play as shown in Fig. 7 at A; also at D Fig. 2. This piece of wire will be very close to the size required, and may be sent to the dealer; if we must order from the supply house, it would be well to order two or three, as the inside diameters will vary, and by so doing, we are quite sure of getting one that will fit.

There is nothing difficult about taking the measurements for a new cylinder. There are three important ones: First, the whole length which is found by getting the distance from

THE CYLINDER ESCAPEMENT.

the outside of the lower cap jewel to the outside of the upper cap jewel, then deducting the thickness of the two cap jewels; this will give the distance from inside of one to the inside of the other cap jewel, or the whole length of the cylinder, allowing nothing for end shake, which should be just enough for freedom. Second, the height of the balance wheel; there are several ways of getting this, but the simplest, and I think the most accurate is to place the hair-spring in position on the balance cock, then measure from the top hole jewel to the under side of the hair-spring collet, being very particular

Fig. 6.

that the spring is flat with the balance cock, or in other words, it is in its normal position, that it will occupy on the wheel. To this measurement, we will add the thickness of the balance arm in the center, and we have the length from the end of the top pivot to the balance shoulder. Third, the most important measurement is the height of the lower slot in the cylinder, the part that is cut away three-fourths of the diameter of the shell. If this is located too high, the watch will not run. If it is located too low, the watch may run in one position and not in the opposite, as when too low the rim of the wheel that supports the teeth would rub on the top of the lower opening and could not enter it freely. To locate this opening properly, we will get the distance from the outside of the lower balance jewel (or cap jewel when we must take off its thickness) to the top of the web or rim of the wheel; to this, we will add one or two-tenths of a millimeter; if the opening in the cylinder is quite large, we may add two-tenths, or if rather small one-tenth will be ample. This gives us the distance from the end of the bottom pivot to the top of the lower opening in the cylinder.

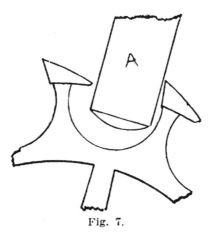

Fig. 7.

When these three measurements are made, we are ready to fit our new cylinder. It is a good idea to first measure up the new one, and see if the brass collet is high or low enough, as it can be driven up or down now more easily than after being once cemented up. We should also get the distance from the top of the lower opening to the end of the bottom plug in order to tell how much to cut off after placing it in the cement. We must use cement in turning our cylinders, as the shell is so hard and so easily broken, we can support it in no other way. Some drive the shell out of the brass and turn the pivot in an ordinary wire chuck, but I do not like that method. Cement is one of the best friends a watchmaker has, and he should be thoroughly acquainted with its use. The plugs are seldom true with the cylinder, and for

188

that reason, we should turn the lower end first and true from the outside of the shell of the cylinder. If we true from the shell, then turn the lower pivot, both will be true with each other, while if we should true from the plug, and it is not true, then the cylinder would not run true with the pivot. After the lower pivot has been finished, the cylinder is taken out of the cement, its length should be carefully measured and then replaced in the cement, this time truing from the end of the top plug, as this is the point that entered the center of the cement chuck before; when this is done, the two pivots must be in line, which might not be so should it be trued up at any other point. By taking the required length of the cylinder from the complete length of the unfinished cylinder, we know at once how much must be cut from the projecting end to make our cylinder the correct length. When this is cut off, we have the top end of the upper pivot, and the other measurements may be taken from this point.

It has been taken for granted that the reader understands the centering of a cement chuck, as this was clearly explained while turning the balance staff, the same principles hold good here.

Great care is necessary that all the cement should be removed from the inside of the cylinder. After boiling in alcohol to remove the cement, it is a good idea to use a little fresh alcohol and carefully clean the outside and inside of the cylinder, as when the cement is dissolved it leaves a thin film on the surface, which appears to destroy the polish.

On the outer edge of the balance wheel, we find a pin projecting. This is the banking pin. We will also find another pin usually on the under side of the balance cock that the banking pin will strike, as the balance wheel revolves; it is very important that the balance wheel be staked upon the cylinder in the correct position on account of these pins. When the cylinder is at rest, the banking pin in the balance should be exactly opposite the banking in the watch, and when the cylinder is at rest, the large opening (the upper one) should be toward the escape wheel, so a line drawn through the two lips of the cylinder would form a right-angle to the line of centers from the balance to the escape wheel.

When in beat, the hair-spring should be so placed on the

balance that the cylinder will be in this same position when at rest.

The use of some parts of a cylinder movement does not seem to be clearly understood; the chariot is one of them. This is a separate piece from the bottom plate in which the

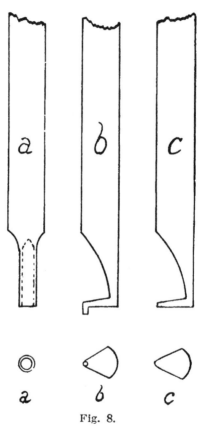

Fig. 8.

lower balance jewel is set. The steady pins and screw of the balance cock, also enter it and by this means we are able to set the cylinder nearer or farther from the escape wheel, and by so doing, can increase or decrease the apparent lift of the cylinder, or more accurately speaking, we will increase the amount of lock, the lift being determined by the height of the incline on the tooth of the escape wheel. It is very important that the lock should be enough and at the same time, it should not be too great. The correct amount of lock and lift is shown on nearly every movement by either two or three dots on the bottom plate near the rim of the balance. The space between the two outside dots represents the distance the balance should move, while a tooth is unlocking and giving impulses to either lip of the cylinder, in other words, the balance should move this distance while the escape wheel makes two successive drops. If the balance

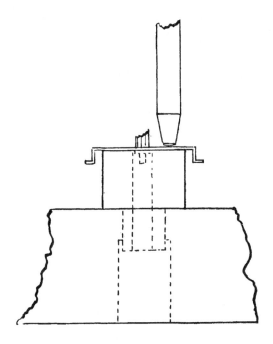

Fig. 9.

moves a greater distance than the space between the two outside dots, the cylinder is set too close to the escape wheel and the chariot should be moved back, and if it should not move so far the chariot should be brought forward until it has the correct amount. When the balance makes a very quick short vibration, the cause is nearly always too light locking, or none at all. We often find a movement where the impulse is good with the exception of three or four teeth, where the quick vibration will occur. This is nearly always caused by the escape wheel being out of center, the fault being a pinion that is not true or the hole in the wheel not in the center. The dots are shown plainly in Fig. 6. In many movements a dot is placed on the balance rim. This dot should be opposite the center dot when at rest, and is sometimes used in putting a watch in beat, although in many instances, it cannot be depended upon, as when a new cylinder has been put in, the balance is often staked on in a little different position from the original; this would not seriously affect the running of the watch, but would be enough to affect the beat.

We often find after a cylinder has been in use for some time that the impulse faces of the teeth have worn little notches in both lips of the cylinder; when this is found the best plan would be to put in a new cylinder, but as our customer will not always be willing to stand the expense, we may repair the old one, so it will give very good satisfaction by grinding out the worn places with a small iron grinder and oil-stone powder and oil, rounding off the corners again as when new, then polishing with diamantine and oil. When this has been done, it will be necessary to move the chariot a little in order to bring the cylinder closer to the escape wheel.

Again, we often find the impulse faces of the escape wheel either poorly polished or rough from wear. To polish all of these teeth would at first seem to be a tedious task, but it can be quickly accomplished in a very simple manner. Place the escape wheel in the lathe by clamping the leaves of the pinion very lightly in a chuck. Now, take a piece of very thin sheet brass or an old main spring from which the temper has been drawn, and file it very thin, place a little diamantine and oil on the end of this piece of metal and hold lightly against the impulse face of a tooth. Revolve the lathe slowly, and the spring in the polisher will cause it to drop from the heel of one tooth on to the point of the next, when it will slide along the impulse face and drop upon the one following, etc. One who has never tried this method, will be surprised to see how quickly it can be done, and also how well.

We often wish to remove one of the plugs in order to put in a new one, and find them hard to start; if it is the lower one, tap the cylinder lightly with a hammer on the outside, which will stretch the shell, when the plug may be easily removed. The upper plug may be treated in the same manner, but it would be necessary to drive the cylinder out of the brass collet. This may be accomplished by using a punch like the one shown at a, in Fig. 8, which will have to be made, as they are not found in any of the staking tools. It will be necessary to have several sizes. The punch at the end should be slightly smaller than the outside diameter of the cylinder so it will rest on the end of the shell, and drive the cylinder out of the collet without affecting the latter. We should have two or three cylinder punches like b, Fig. 8. These may be found in all staking tools, but are too large for use, and as many or most of the cylinders are very small, we will be required to make our own punches. These should be tempered to a blue; a punch of this kind is used for driving out the plugs; another form c, same illustration, is very useful for driving the cylinder up when necessary to lower the brass.

When an escape wheel is not true in the flat, it is liable to strike either the top or the bottom of the opening. If it touches on the bottom the watch would stop, and if on the top, the wheel would recoil when the lower end of the top lip struck it. The steel wheels are very hard and difficult to bend; the fault may be caused by the wheel not being staked on the pinion properly, but many times the wheel has been bent. We may true a wheel by bending the arms, which may be done by turning up a small brass stump to fit our bench block and using a round face punch, as shown in Fig. 9; a hole is drilled in the center of the stump large enough to allow the pinion to enter; if we wished to raise one side of the wheel, we place the punch on top of the arm. If we wish to lower one side, we place the wheel bottom up and the punch on the under side of the arm, being now of course the upper side. By using a brass stump (a lead one is sometimes used), the metal is soft enough that the hard steel will bend and there is very little danger of breaking a wheel.

The spaces between the teeth of an escape wheel are seldom all the same, as all steel will spring more or less in hardening. When the spaces are irregular, it is sometimes necessary to grind a trifle off from the back of some teeth that the cylinder may be free all around.

THE DUPLEX
ESCAPEMENT.

The Duplex Escapement was one of the first invented, and for a time was in general use, but the Lever Escapement was so much superior, that it took the place of the Duplex and for a number of years the manufacture of the latter was discontinued until one of the American manufacturers brought out a cheap watch containing this escapement and today they are rapidly taking the place of the cheap cylinder watches of foreign make. This enterprising company makes a very attractive line of cases and a great variety of movements, and for a cheap watch they are giving very good satisfaction.

It seems almost out of place to say much about the repairing of the Duplex Escapement at this time, but as they are now in general use, and we are called upon to clean and repair them, it is quite necessary to understand the principles upon which they are constructed.

In the Lever Escapement, the balance is free from the lever and the train during the greater portion of its movement, and for that reason it is called the "Detached Lever." In the Duplex, the escape wheel teeth are in contact with the balance staff nearly the whole time. It is often spoken of as a "dead beat," but is not, strictly speaking, as the escape wheel recoils as we will presently see.

The duplex escape wheel is a *double one* as it has two sets of teeth, one for locking and the other for impulse. At first, two wheels were used being staked upon the same pinion, but this was found to be less satisfactory than having all of the teeth cut from a solid piece of metal. There have been many changes in the shapes of the teeth and the style of the wheels. Many of the old escapements had wheels of the form shown in Fig. 1; here we have double teeth for locking and a single tooth for impulse. We have twelve locking teeth (six double ones). Where a wheel of this form is used the balance will receive an impulse only every *fourth*

195

vibration as when the first tooth passes through the notch in the staff or locking roller, the second tooth locks against the staff or roller, and when the second tooth passes through the notch, the impulse arm should be in such a position that

Fig. 1.

the impulse tooth will drop upon it, and give the balance its impulse. The amount of drop is quite an important thing. We should always keep in mind while working on an escapement of this kind, that if we increase our drop, we must decrease our lift or impulse, and if we decrease our drop, we will increase our impulse, and consequently the motion of the balance. It will be easily understood that we should give as little drop as we can, and have the action of the escapement perfectly safe. Fig. 2 shows the style of escape wheel used in the Duplex Escapement now being manufactured. A single tooth takes the place of the double ones shown in Fig. 1. In this case, the balance will receive an impulse each

Fig. 2.

alternate vibration. An escape wheel of this kind has as many locking teeth as impulse teeth.

The action of the escapement may be better explained by the aid of the drawing shown in Fig. 3. A represents the escape wheel a a a the locking teeth, b b b the impulse teeth, which project above the top surface of the rim of wheel; c is the locking roller, or, as it is made at the present time, is the lower end of the staff, the notch d being milled out for the

197

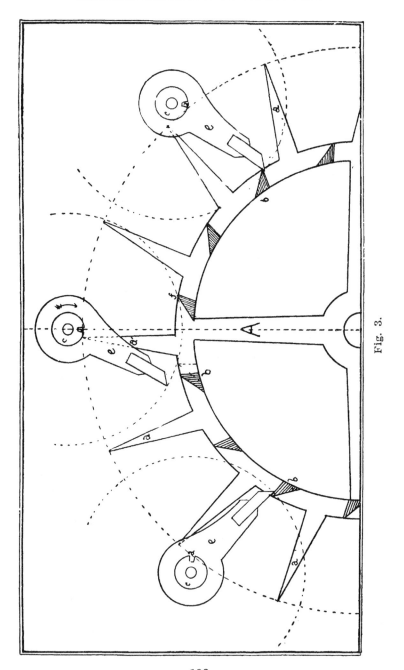

Fig. 3.

tooth to pass through; e e e is the impulse arm. It is usually staked upon the staff, and is moved forward or backward to regulate the drop. In some of the movements instead of this impulse arm, a pin projects from the underside of the balance arm, the impulse teeth of the wheel drop upon this, giving the balance impulse in the same manner as before. It will be seen that the locking teeth lock but a very small amount on the roller c, so that a very little side shake of the pivots of either the balance staff or escape pinion would greatly affect its action, and for this reason it is necessary that the pivots should fit the jewel holes more closely than in any of the other escapements. The action of the escapement is as follows: the tooth a is locking against the surface of the roller c, or against the staff, as the case may be. The balance moves in the direction of the arrow; as soon as the notch moves around far enough the tooth drops into it, but is immediately thrown out as the notch passes by, the elasticity of the hairspring brings the balance back and the tooth a drops into the notch d again, and this time passes forward with it until it passes out on the opposite side. If there was nothing to interfere the tooth back of this one would immediately pass forward and occupy the position the first one had at the start, but we find something to prevent this. At the moment the tooth a passes out of the notch in this small roller, the impulse arm has moved forward until it is just in front of the tooth b, so that when the wheel moves forward, this impulse tooth drops upon the end of the impulse arm e, giving the balance impulse until it passes off from it at the point shown by the intersection of the dotted circle at f. When this tooth leaves the impulse arm, the locking tooth is near the roller and the real drop is but a small amount as impulse has been given the balance while the wheel has been moving forward, these movements are repeated during each double vibration. At the left of the drawing, the impulse arm is shown in the position it is in when the wheel drops or at the *beginning* of the impulse. At the right side of the same drawing the wheel and impulse arm are shown in their position, at the end of the impulse or just as the wheel is about to drop and the long tooth lock on the roller. The impulse arm, roller and escape wheel are shown in the center as they would be when in a state of rest, or when the Duplex Escapement would be in beat, it will be noticed that the notch in the smaller roller is on the line of centers denoted by the dotted line. There are some cases where the watch will give better satisfaction when the notch is a trifle in advance or back of this position, yet in most cases, it is placed on the line.

The drop may be changed by moving the impulse arm on the staff, where it is held by friction. If we bring the impulse arm and the notch in the roller near together, we will increase the drop, as we increase the drop, we decrease the impulse, so we should give as little drop as possible, and have it safe. If we move the impulse arm in the opposite direction, denoted by the arrow, we will decrease the drop, and of course, increase the lift.

Two lifts or impulses are given, the primary and the secondary, the latter is given as the tooth passes through the notch in the roller and is not a very powerful one, yet enough to be easily seen. The primary or main impulse is given as the upright teeth drop upon the impulse arm, the length of this arm being longer and the radius of the wheel shorter, the balance will receive a very powerful impulse.

A Duplex Escapement when properly set and having no hair spring, will impel the balance so it will revolve continuously in one direction until the watch runs down.

In the oldest movements the locking roller was made of either ruby or garnet cemented to the staff, the surface of the jewel being highly polished reduced the friction of the locking tooth. The impulse arm was also jeweled so the friction was reduced to the smallest amount possible.

A word about selecting a roller of the correct size may be helpful. If we examine the action carefully, we will notice that the end of the impulse arm as it returns after receiving its impulse just passes the upright tooth in the wheel, the one nearest the center at the right in the drawing, and as the arm returns from the other side the end will just pass the tooth at the left of the center denoted by the dotted lines, the wheel moves forward as the locking tooth passes through the notch in the roller and is in the position denoted just at the drop. The action just explained is the correct one. If the end of the impulse arm should barely pass the tooth on the right and have considerable space as it returns on the left, this would show at once that the roller was too large. If on the contrary, the space at the right was large and at the left very little, then the roller would be too small.

Should it be found necessary at any time to make a new impulse arm, its length may be determined as above, making it as long as possible and pass the teeth on each side.

The American made duplex watches have advantages over those made at first: Where the locking roller is separate from the staff, the notch in the roller must be quite shallow, consequently the locking must be very light, and the friction or point of rest being nearer the line of centers would have a tendency to retard the motion, while the later ones are made with the locking roller, a part of the staff, the notch being milled in the side and this notch often is more than half way through the staff. This permits a greater locking, making it safer and with less friction. A staff to fit any of the movements can be bought for a few cents, but often we will lose the job unless it can be done in a short time, so in a case of emergency we ought to know how to make one. The most difficult part will be to make the notch for the tooth to pass through. If one has a wheel cutter or milling attachment, it can easily be done, but as most workmen do not possess these, a simpler and just as effective a method will be given. We may turn down a piece of steel wire which has not been hardened, until it is a little larger than the lower end of the staff is to be when finished. Now, make a steel punch with the end the shape the notch is in the staff, harden and temper to a straw color, then place the turned part of the wire on a bench block or some hard surface, and drive this punch into the soft steel where required. It may spring the wire and make it out of true, but we may put the wire in the lathe, and by bending make it run true, when it can be hardened and tempered to a blue, and the staff finished. I have made a great many in this manner that gave good service.

It is quite important at the present time to be able to do this kind of repairing, as these watches are coming into very general use, and are rapidly taking the place of the cheap cylinder movements that are so poorly made.

THE CHRONOMETER ESCAPEMENT.

The average watchmaker is less familiar with the chronometer escapement than any in common use; in some localities he is never called upon to clean or repair them while in other places they are much more common.

Pocket chronometers never came into general use on account of their delicate construction, making them unsuitable for portable time pieces and not giving the satisfaction of a good lever escapement, which is less expensive.

For marine time pieces, the chronometer surpasses all other escapements, and "Marine Chronometers" are used by the best jewelers throughout the country to denote their standard time, and to regulate their watches.

A ship at sea must determine its location by the time of its chronometers, a variation of only a few seconds making a difference of many miles, so it will be seen at once they require the most accurate time piece that can be made. As a safeguard, the large ocean liners and the great battle-ships have three or four fine chronometers, each having a very close rate. By comparing them with each other they can take the average of them all, and in this manner obtain more accurate time than could be possible with only a single one. In the Naval Observatory at Washington, all of the chronometers used in the navy and other government vessels are rated; one room is devoted to this work; all new ones are also rated there before sending out. A perfect time piece has not yet been made, and the finest chronometers have either a constant gaining or losing rate, and this gain or loss must be deducted or added when determining the longitude at sea.

The chronometer, while of delicate construction, is at the same time one of the most simple we have to deal with. It is often classed with the dead-beat escapements, but as the locking jewel in the detent is set at such an angle that it will have draft, and in all cases where there is draft, the escape wheel must recoil in unlocking, the chronometer cannot be a dead-beat, although it is the most free or detached of any escapement now in use.

Marine chronometers have one advantage over all portable time pieces—they have to be adjusted for *one position only*, as it is suspended horizontally in gimbals, and the outer case of the movement is often heavily weighted in order that it may always remain in the same position. The great advantage of having a movement constantly in the same position can be readily seen.

In these days of twenty-one and twenty-three jeweled watches, one would be inclined to think a chronometer,

Fig. 1.

which has such a fine rate must have many more. Such is not the case, however, as many of the very finest ones have but nine or eleven jewels. A good, hard brass bearing for a pivot is about as good as a jewel if well oiled, and will wear a great many years.

By keeping the movement constantly in one position, we have reduced the position adjustment to the minimum, and by the construction of the movement we may also reduce the isochronal adjustment. This is done by the aid of the fusee and chain, which is made like the ones found in English lever watches, only larger. When the mainspring is fully wound, we know it has a stronger pull than when nearly run down. In an ordinary watch, the balance will have a longer arc of vibration when first wound than when nearly run down. The fusee and chain prevents this, and if properly constructed, the arc of vibration will be the same at all times. Fig. 1 shows us the fusee, chain and barrel of a chronometer. When the spring is fully wound, the chain acts upon the small part of the fusee; the leverage being small, this counteracts the extra power of the fully wound spring. When the spring is nearly run down, the chain is pulling on the largest part of the fusee, and consequently exerts more power; so when a fusee is properly graduated, the force of the main spring is transmitted to the train of the chronometer without any perceptible variation, and if this force is not variable, the balance will have the same arc of vibration at all times, and if so, will make them in practically the same time. This

Fig. 2.

being the case, the isochronal adjustment can be easily accomplished.

Fig. 2 shows a general view of a marine chronometer. The chain is nearly all wound around the barrel, showing

205

it is about run down. The balance and the cylindrical spring are seen above the top plate.

Another thing where the chronometer differs from most time pieces is, the absence of a regulator. This is better shown in Fig. 3, which is a top view and shows the balance, with its weights, the balance cock, with the hair-spring stud,

Fig. 3.

but no regulator of any kind. It has often been said that a watch that was fully adjusted should have no regulator, as the movement of the regulator would change the length of the spring, and, consequently, its isochronal adjustment. This fault is entirely eliminated by the construction of the balance used in the chronometer. The one here shown (Fig. 4) has the auxiliary compensation, the two screws at the end of the arms are used in timing, being turned out to make it run slower and toward the center to make it run faster. When the chronometer has a very close rate, it is almost impossible to turn the screws little enough to bring it to time. The two round weights are used instead of screws, and are made to slide freely on the rim of the balance. This is better shown in Fig. 5, which is a photograph of the underside of the balance. The weights are held

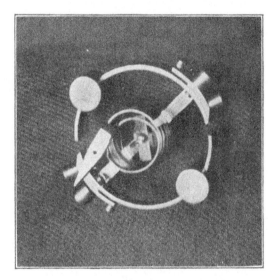

Fig. 4.

in place by small screws on the inside bearing against the steel part of the rim (not shown), as these weights must be moved nearer or farther from the end of the rim, as required for the temperature adjustments (which was explained under the subject of compensating balances and pendulums). It will be seen that we are able to get a finer adjustment by the sliding weights than could be possible with screws which must be moved from one hole to the next, when they might require only a part of that distance to produce the required effect.

The escape wheel must be made very light that it may move quickly and impart the impulse to the balance without any loss of time. The teeth are made quite thick, but the rim of the wheel is very delicate. Fig. 6 shows such an escape wheel. The teeth are pointed and resemble those of an English lever watch.

The most delicate part of the escapement is the detent. This is made of two forms, the pivoted detent where the detent is supported on an arbor that is pivoted at each end, and the tension is given by a small hair-spring on the arbor. The tension can be increased or diminished by turning the collet on the arbor. This form is used in pocket chronometers. The spring detent is shown in Fig. 7. The detent and spring are made of one piece, the spring being ground very thin. The photograph shows a side view. The gold spring is also shown, held in place by the screw. The jewel, the teeth of the escape wheel lock on, may also be

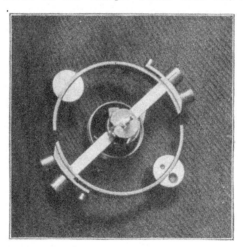

Fig. 5.

seen projecting above the top of the detent. A better idea of the appearance of the top, the detent and gold spring may be had from the drawing, Fig. 8. The action of the escapement may also be better understood from this drawing. It consists of the following important parts:

The escape wheel A, the impulse roller B, the unlocking pallet or roller C, the detent D, and the gold spring E.

It will be noticed that the gold spring rests against the

208

detent at the end, so while the balance turns in one direction, the unlocking pallet moves the gold spring away from the detent and immediately it comes back against it again as

Fig. 6.

soon as released, but the detent has not been moved. Now, when the balance turns in the other direction and the unlocking pallet comes in contact with the gold spring, it carries the detent with it, and when the detent moves far enough to allow the tooth of the escape wheel, which was locking on the jewel in the detent, to pass off, the wheel is free to drop, but as the rollers have moved forward, the jewel (F) in the impulse roller has also moved forward until it is just in front of a tooth of the escape wheel, and this tooth drops upon the jewel in the roller, giving the balance impulse until the tooth passes off from the jewel. Meanwhile the unlocking pallet (G) has released the gold spring and the detent has gone back to its original position so the tooth of the escape wheel will lock on the jewel in the detent again. The drop should be as little as we can give it and have it safe. The less drop we give it, the greater the impulse.

To have the escapement work properly, the unlocking pallet should move the detent forward until the escape wheel drops upon the impulse jewel in the large roller. Immediately after the drop, the gold spring should be released and detent come back to its former position, leaving everything perfectly free to give the balance its full impulse. A

209

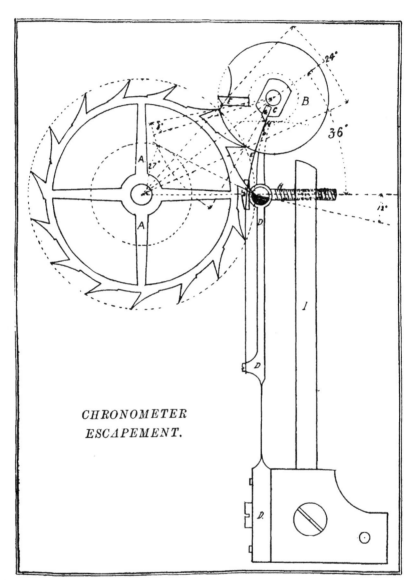

CHRONOMETER
ESCAPEMENT.

Fig. 7.

210

THE CHRONOMETER ESCAPEMENT.

very common fault is to have the gold spring too long, so it will not release the detent soon enough. When this is so, the force of the train is imparting impulse to the balance, while the force of this spring is holding it back, and one force counteracts that of the other to a certain extent.

The space between the teeth and the circumference of the large roller should be about the same. The points of the teeth should not come in contact with the outside of the roller. If these spaces are unequal, they may be corrected

Fig. 8.

by setting the detent forward or backward. By moving the detent forward, the space between the tooth will be diminished on the side toward the detent, and the space on the opposite side will be increased. The space in front should be a trifle greater than that at the back to allow for the recoil of the escape wheel in unlocking.

If we move the two jewels in the rollers nearer together, we will increase our drop and decrease our lift or impulse.

A chronometer escapement will not start itself when the power is applied like a lever watch, but must be started by moving the balance enough to unlock the tooth from the locking jewel of the detent.

A chronometer should be put in beat by placing the hair-spring on the staff in such a position that when the balance is at rest, the unlocking jewel will be just in front of the gold spring and resting against it. The position of the rollers and detent in the drawing is at rest.

211

The banking screw, H, is used for adjusting the lock. The detent rests against the head of this screw. By unscrewing it, we will allow the detent to come closer to the escape wheel, giving more lock, and by screwing it in, the detent will be carried away from the escape wheel, giving less lock.

Most marine chronometers have 14,400 vibrations per hour, or 240 per minute. As the balance receives an impulse only each alternate vibration, it would receive 120 per minute, or two a second. The second-hand moves each time the balance is given an impulse, so the second-hand will move twice every second, or once every half second.

Most chronometers are made to run fifty-six hours, but are usually wound every twenty-four hours.

The average workman is not often called upon to repair chronometers, yet he should be familiar with their construction, as it requires even more skill to do the work that is not common, than it does to do that which presents itself day after day. The chronometer is very delicate, and requires carefulness, otherwise it is not so difficult to clean or repair.

CLEANING AND OILING.

Human ingenuity has conceived and constructed many ponderous machines and delicate instruments that are almost life-like in their actions. None of them are more delicate or are expected to perform better, or for so long a time without attention as our time pieces. An ordinary machine which is in use only eight or ten hours each day is carefully oiled one or more times daily, and yet the average time piece is expected to run from one to three years after being cleaned and oiled. When one knows the oil is constantly thickening and is exposed to the heat of Summer or intense cold of Winter, and during all of these changes and conditions the time pieces are not expected to vary but a few seconds for months, a very difficult problem confronts the watchmaker.

It is generally conceded that anyone can clean a watch if he can do anything, as this is one of the first things an apprentice is taught to do; yet, I claim it takes as much skill to clean and oil a watch properly as anything we are required to do. Some workmen brag of their speed, claiming to be able to clean a watch in twenty minutes. No one can do it properly in that time. Of course, each individual has his own method of doing the work, which he thinks is the quickest and best. Speed in these days of strong competition is an important factor, but quality of work should never be sacrificed in order to gain speed.

One of the most important things we often neglect is the proper inspection of the work before setting a price on the repairing. If the following plan is carried out, you can get one-half more for your work and have a better satisfied customer as well—two very necessary requirements. When a watch is brought in for repairs, don't glance at it and say it needs cleaning and will cost $1.50, as too many are in the habit of doing, but tell your customer to call again in an hour or two, as your time demands, or if he cannot call again, take time to properly examine it while he is waiting, take the watch down carefully, examine the pivots to see if they require polishing, examine the jewels, some may be broken; the stem wind parts should be well inspected; any of these defects show your customer, allowing him to look through your eye glass, show him a pivot that is broken and explain that you must drill a hole in that piece, carefully fit a piece of steel to it and turn a new pivot like the other one. After showing what must be done, he will feel satisfied to pay a good price, for he knows it will require great skill to do such delicate work; the same person would be very much dissatisfied to pay less if he did not see what had to be done.

If you look over a watch hastily and tell your customer it will cost him $1.50 to do the work, you are in duty bound to do it for that if it costs you more, for it is your fault that you did not know what was required, and if you try to convince your customer it is worth $2.50 when you said it would cost only $1.50, you will have a hard time of it, and if you do succeed, there are liable to be hard feelings.

It is a good plan to have some small wooden boxes, like Fig. 1, made to keep the movements in when taken down for inspection, also some little pieces of paper on which may be written the items of repairs necessary; then when your customer returns, the movement is in shape to show him any defects and the record is with it on the paper.

It will seem at first that this method will require too much time, but all work must be examined before the repairing is done and the extra amount that can be charged for the same work will more than pay for the small amount of extra labor, and what is more, you have a satisfied customer, the best of it all.

By this method we have a list of what repairs are necessary when ready to do the work. By checking these off as completed we know when we are ready to clean the movement. Too many begin the cleaning before the repairing has been done, with the result that some parts will have to be cleaned several times. We should make it a rule to always make any needed repairs, polish any pivots that are black or rough, and place every part in first class condition before beginning the operation of cleaning.

Every watchmaker has his own method of cleaning. Of course, he thinks his way is the best. Some have gone so far as to invent machines for doing this work. I am satisfied that most of the work must be done by hand.

We should procure some good benzine (not gasoline), only the very best quality being suitable for watch work;

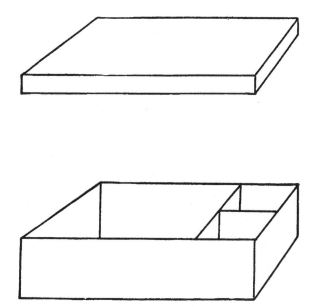

Fig. 1.

some grain alcohol, some cyanide of potassium (a deadly poison), and a cake of pure castile soap (Ivory soap may be used, but it is not as good).

We will make a solution of the cyanide of potassium by dissolving 1 to 1½ ounces of it in a quart of water, rain water being the best. Many form a wrong idea of the use of this cyanide solution. It does not clean off the grease or dirt, but only brightens the surfaces and gives them a newness that is desirable. This solution should be labeled

215

poison and only used for watch work. It should also be kept in a covered receptacle.

The alcohol is used to dry the parts rapidly and to prevent any parts from rusting that should not be properly dried.

Our method of procedure is about as follows: String the larger parts, like plates, bridges and stem-wind parts, on small wires, having a loop on one end, slip the other end of the wire through this loop and the parts are secure. On another similar wire, place the train wheels and smaller parts; the escape wheel is placed on another one by itself in order to prevent any damage to its delicate teeth.

If any of the parts are very greasy or the oil has become very thick on them, they may be placed in the benzine a few moments; this will cut the grease and gum; if not in bad condition, this operation may be dispensed with, and the parts carefully washed by using a medium stiff brush and soap and water, using soft water when it can be had; this removes the dirt, after which it is well rinsed in water. Now, we place it in the solution of cyanide for only a moment, which brightens the parts, when it should be thoroughly rinsed again to remove all of the cyanide. The parts are now placed in alcohol, which absorbs the water; after remaining in the alcohol about a half of a minute, they should be placed in fine box-wood sawdust and kept in motion until none of the sawdust adheres to the parts, when we know they are dry.

Parts that are cemented in with shellac, such as the pallet stones and jewel pin, should not remain in the alcohol, as the cement would be dissolved. They may be dipped in the alcohol and immediately dried in the sawdust without any danger, but in most cases it is not necessary.

All cap jewels or end stones must be removed in cleaning. Many workmen remove them all before cleaning, but by so doing they sometimes become changed or the hole jewels may get transposed and cause considerable trouble, especially if the escapement is cap-jeweled, so instead of taking them out before cleaning, I always leave them in the plates and also in the balance cock and potance until these parts are cleaned, then remove one of them, thoroughly clean the cap jewel and peg out the hole jewel until it is also perfectly clean and replace the cap jewel and the screws, proceed the same with each of the cap and hole jewels. By so doing, there is no possible chance of making a mistake, and it can be done in less time. All of the pivot holes and jewels should be pegged out until the pegwood remains white when taken out. It is also a good idea to go through each leaf of the pinion with the point of a pegwood, as there is always a thin film on the surface of the steel that can hardly be removed in any other manner.

It is necessary to brush out all parts with a soft, clean brush in order to remove every particle of sawdust. In brushing the plates or any finished part, always brush in a circular direction, otherwise the parts are liable to show the marks left by the brush.

From the time the parts leave the sawdust, they should not come in contact with the fingers in any way, as nothing is so unsightly as a watch plate with finger marks upon it. This is not a mere matter of looks, but if a finger mark should be left on a piece of polished steel, in a short time it would rust, caused by the perspiration, and nothing will remove it except refinishing the surface of the steel, which in many cases would ruin it. All parts should be handled with the tweezers and the plates held in the best tissue paper that can be procured, that of Dennison's being about the best.

The lever and the balance pivots may be cleaned with a soft piece of pith, or they may be placed in a solution of equal parts of benzine and sulphuric ether. This will clean off the old oil quickly and the surface dries at once, as the benzine and ether evaporate so rapidly. This solution is a splendid one for cleaning the hair-spring, and is superior to benzine alone in all cases, but more expensive. It should not be used near a flame.

There is a difference of opinion about removing the main-spring from the barrel in cleaning. The spring is the life of the watch and must be in good condition, and if the oil is thick and gummy, I always remove it carefully from the barrel, but never put it in benzine to clean. If it is very gummy, place plenty of oil on it or dip it in sperm oil;

this will loosen all of the old oil, and we may now carefully wipe off the oil with a soft, clean cloth or a piece of tissue paper, when the spring may be replaced with a winder. There is no great danger of breakage when done in this manner, but when the strain of the spring is suddenly released in removing from the barrel, there is danger of its breaking soon after being replaced.

All parts are now supposed to be thoroughly cleaned, and we are ready to put the movement together and oil it.

Many very good workmen are quite careless about oiling. This important part requires as great care and should be

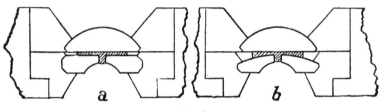

Fig. 2.

done as carefully as any of the cleaning. The stem-wind parts should be oiled as they are replaced, clock oil being better than watch oil, as it is a trifle thicker. None of the teeth of the wheels should be oiled, except the ratchet teeth found in many Swiss and some of the later American movements. The ratchet teeth and the square winding arbor upon which one of the wheels slide should be oiled. I have found many cases where the winding seemed very rough, particularly when turned backward, that would be very free and easy after being properly oiled.

The main-spring should be thoroughly oiled as the coils slide upon each other, and we must reduce the friction as much as we possibly can.

A very important part is the pivot of the center wheel that passes through the pillar plate. We often find this pivot perfectly dry and badly cut from not being well oiled. There are cases where the cannon pinion comes too close to the plate and the capillary attraction draws the oil away from the pivot, after which it soon begins to wear. The cannon pinion should never come in contact with the plate, the shoulder of the pivot being long enough to prevent it. In oiling this it is a very good idea to place a small amount of oil on the shoulder of the pivot before placing it in the plate, also oiling again afterward.

There is a great difference of opinion about the proper method of oiling the balance pivots or any jewels having end stones or cap jewels. Some place the oil on the flat surface of the jewel before replacing it. This, I think, is the very worst practice possible. Take, for instance, the regulator cap of a Swiss watch. When oiled in that manner the oil will be smeared all over the top of the balance cock

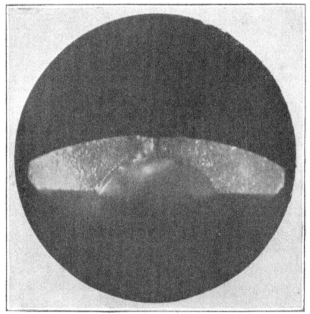

Fig. 3.

before the regulator can be put on and the screws replaced. The capillary attraction will take the oil away from the pivot instead of drawing it to it.

The best method is to carefully clean the jewels and replace them, not oiling until later, as will be explained.

The surface of the balance jewel next to the cap jewel was at first made perfectly flat, but the best jewels now in use have a rounded surface, which has a decided advantage in oiling. This may be more clearly seen in Fig. 2 at a. We have a flat cap jewel and a flat balance jewel. These jewels should never touch each other, but should be so set that their surfaces leave a small space between them. It will be seen by the shaded portion the position the oil occupies; at b, we have a flat cap jewel and a rounded balance jewel, the shaded portion showing the oil here also. In the first case, the pivot may be well oiled, but there is nothing that

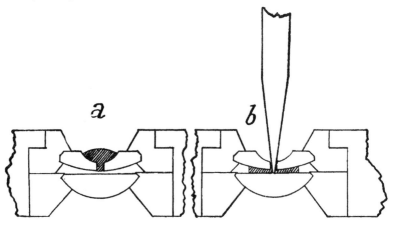

Fig. 4.

will bring the oil to the pivot; in the second case, the oil at the end of the pivot is replaced with fresh oil by capillary attraction as rapidly as used, and the oil will continue to be brought to the pivot as long as any remains between the jewels. Fig. 3 is a photograph of a modern balance jewel.

Now, let us consider the best method of oiling these jewels. An objection was made to oiling the cap jewel before replacing it, and we cannot always depend upon the oil passing through the hole in the balance jewel and reaching the cap jewel, even after the pivot enters. The following plan is one of the very best, you are sure of the oil being in just the right place, and a watch so oiled will run from six months to a year longer than one where the oil is placed on the cap jewels before they are replaced. First place a small amount of oil in the cup of the jewel as shown in Fig. 4 at a. Then take a very sharp piece of pegwood and place in the hole of the jewel, allowing it to pass through and touch the cap jewel. This carries the oil through, and

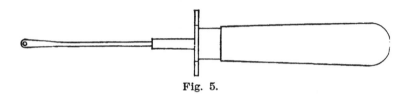

Fig. 5.

the moment the pegwood comes in contact with the cap jewel the oil will disappear from the cup, and it may be distinctly seen between the two jewels. We may now place some more in the cup of the jewel. I have seen several cases where oiled in this manner that the center of the cap jewel would

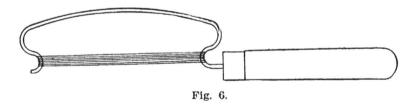

Fig. 6.

resemble a series of concentric rings, those near the outside being almost transparent, and growing darker and darker toward the center, thus showing the condition of the oil as it grew old and thicker.

There are a great many kinds of oilers, some like the fountain oilers, some use pegwood, others the small gold ones, etc., but none of them seem to quite fill the bill. The one shown in Fig. 5 comes the nearest to perfection of any I have yet used. We have an ordinary gold oiler with a very small hole drilled through near the end. The hole should be about one-tenth of a millimeter, and the metal around the hole should be just enough to give the necessary strength.

221

It will be seen at once that the oil will be held at the end of the wire by the capillary attraction and the moment the oiler touches a jewel or a pivot the oil at once enters the jewel, or remains on the pivot. I am sure anyone taking the time to construct such an oiler will feel well repaid after using it a very short time.

It is a good plan to often clean the oiler by sticking it into a soft piece of pith.

In oiling the escape wheel and pallets, I have found it best to do it about the last thing, while the watch is running. Touch two or three of the impulse faces of the teeth with the oiler; by so doing the oil will only touch the pallet stones where the teeth pass over them, while if the oil is placed on the pallet stones the whole surface is liable to be covered.

We often find some old movement where the plates are badly tarnished, and the process of cleaning mentioned will not brighten the plates or remove the tarnish. Such obstinate cases may be brought out almost like new by using some fine powdered rouge with the soap and water. It is surprising how their appearance will be changed by so doing. An article that is very helpful is a chamois buff, nothing more than a strip of soft chamois skin glued to a piece of wood. This is coated with rouge and is often useful for brightening up parts or removing tarnish. Another simple article is the balance polisher, which can be easily made by bending a piece of wire and winding linen thread upon it. This is shown in Fig. 6. Rouge is used on the thread. The chamois buff should be kept in a paper case in order to keep it free from dust and dirt, as such particles would scratch the nicely finished surfaces.

I will explain another method of cleaning which gives even better results than the method just explained. To those not familiar with the process they would say it would take too much time, but when everything is ready the work can be done in much less time than they would imagine.

The equipment is a little more elaborate, and would have to be made, as nothing of the kind is now on the market. We will need four quart bowls, either glass or china, with covers on two of them; these may be easily procured.

CLEANING AND OILING.

A copper dish similar to a small dipper holding about a pint, two copper cups with brass sieve bottoms and wire handles. These must be made and should be large enough that the largest size watch plates may be placed in them; to these add a piece of Turkish toweling, about ten by twenty inches in size, some castile soap and powdered borax, and our equipment is complete, supposing, of course, that we have a small gas burner that we can use for heating the water.

We will proceed as follows: Take down the movement carefully, place the plates and larger parts in one of the cups with sieve bottom; in the other one place the wheels and more delicate parts. In my own work I usually clean two watches at one time, either a large and a small one, or an American and Swiss, the idea being to choose those whose parts are not at all alike or liable to be interchangeable.

In bowl No. 1 we have cold water, bowl No. 2 is empty, No. 3 has the cyanide solution, same as before used, and bowl No. 4 has pure grain alcohol. We fill the copper dipper with water and place on the stove. While heating, shave up about a teaspoonful of the castile soap, add to the soap a teaspoonful of the powdered borax. When the water boils, pour one-half of the contents of the dipper into bowl No. 2, then put the soap and borax into the water remaining in the dipper and heat again, watching very closely. As soon as it begins to foam, remove from the fire and take the cup with the watch plates and place in the dipper, and again heat the solution, moving the cup up and down in the solution until thoroughly cleansed; the same thing is done with the wheels and smaller parts in the other cup. The borax and soap removes all of the dirt and makes the parts look like new.

Now place the cups containing the parts in the dish of cold water. This removes all of the soap and borax. Next dip them in the cyanide solution in No. 3 just a moment, after which we place them in the hot water in No. 2. This removes the cyanide, after which we place them in the alcohol in No. 4. The alcohol absorbs the water and the parts may all be quickly dried on the Turkish toweling, which has been spread out on a flat surface, the pieces being placed on one-half of it, and the other half folded lightly over them.

The alcohol is quickly absorbed and everything will be found to be nice and clean. It is hardly necessary to peg out the jewels, but it is better to do so. Of course, by this method we must remove the cap jewels the same as with any other method of cleaning.

Our brushes should always be kept exceedingly clean. We need a hard one for the coarse work and a fine, soft one for the more delicate work. It is a good idea to have two sets of brushes; then as one becomes soiled they may be washed and the other set used while drying.

A new brush is too harsh to use, and should be prepared before using. Some remove the roughness by drawing the bristles back and forth over coarse sand paper. Another way is to pass the sharp edge of a piece of glass back and forth over the bristles; by so doing the brush is rendered very soft and the bristles are more wedge shaped and very thin at the points. When so treated there is no danger of the brushes scratching the finished parts.

WHY SOME WORKMEN FAIL TO SUCCEED.

"To fail at all is to fail utterly."

Why do some workmen fail? Why does their work give poor satisfaction? Why are their customers always complaining?

Is there any reason for such failure, for dissatisfaction, for complaint? Unfortunately there is; many of these reasons will be illustrated by photographs, no better way of showing the work which is the real cause of these failures can be used.

Many a workman thinks because much of his work can not be seen "anything is good enough," as long as the watch will run. That is not honest, your customer pays for good

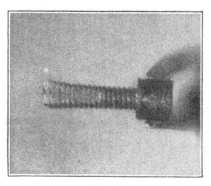

Fig. 1.

work and he should receive it. There are so many little things requiring but a few moments to do properly, that many take a much longer time in doing and then it is poorly done, for instance, the simple matter of fitting a screw. It should take but a few minutes to make one that will fit properly and the work will be well done. I have seen men spend an hour trying to find a ready made one that would fit, and even then it was not a satisfactory job; again, the threads in a plate become "stripped" and the screw will not hold. Often the hole in the plate may be closed or bushed, correcting the trouble. The poor workman will pound the threads with his hammer, flattening the end, making it larger; it may hold for a very short time, but only makes matters worse. Such a screw is shown in Fig. 1, and was in actual use.

We find some very amusing methods used to repair broken parts, one of the queerest that ever came to my

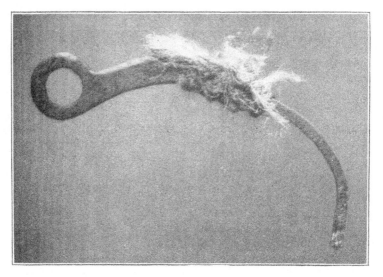

Fig. 2.

notice is that shown in Fig. 2, a click spring for a Swiss watch. It was broken and evidently the workman could not make a new one, so he tied the old parts together with thread allowing the two pieces of metal to lap a trifle.

Sometimes it is necessary to repair certain parts by soldering with soft solder. When this is done, we should be very careful to thoroughly clean it of all acid, otherwise it will soon rust. Fig. 3 shows a broken piece that was poorly repaired by soldering in a piece of steel to replace the broken part, but it was not well cleaned and in a short time it was badly rusted, as plainly seen.

WHY SOME WORKMEN FAIL TO SUCCEED.

Often in examining our escapement, we find the impulse faces of the escape wheel teeth are so high or low that they do not act in the center of the pallet stones; in American

Fig. 3.

watches it would not make so much difference, but in the Swiss, where they have enclosed pallets, the tooth is liable to pass along the steel instead of the jewels; in all cases to

Fig. 4.

have them correct, the center of the impulse faces should be the same height; if the pallets are too low, they may often be corrected by placing a nicely turned washer of the same diameter as the pallet arbor between the shoulder of the arbor

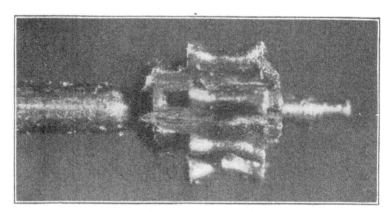

Fig. 5.

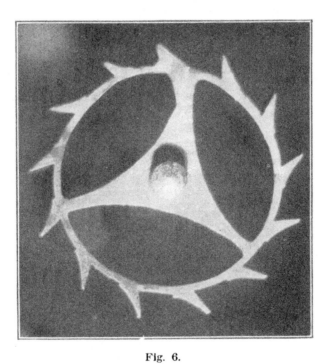

Fig. 6.

and the pallets, the thickness being determined by the amount the pallets should be raised. There can be no objection to such work when well done, in fact, no one should be able to detect it unless very closely observed, but when a man does a piece of work like that shown in Fig. 4 by punching a hole in a piece of main-spring and making an excuse for a washer, which projects beyond the pallet arbor far enough that the teeth of the escape wheel will touch it and then expects a watch to keep time, it is about time for him to retire from the business.

Fig. 5 is a good illustration of pure carelessness or a lack of judgment of the strength of materials. It shows an escape pinion upon which the wheel had been riveted, either the wheel fitted too closely or the punch was struck too hard for the leaves are badly distorted, and yet they tried to obtain time after such treatment. It is needless to say it was not a success.

The escapement is the most important part and requires better treatment and more careful work than anything connected with our time pieces, and yet one would believe after inspecting Fig. 6, such was not the case. Here we have a chronometer escape wheel where a tooth has been inserted, soft soldered in place and filed up by hand. This is an actual case taken out of a marine chronometer supposed to have been used as a standard time piece by a jeweler. Do you wonder that it did not give satisfaction? We expect the finest work that can be done on chronometers, yet some will attempt to obtain a good rate with such poor workmanship as illustrated.

The escapement is the "life of the watch" and requires the greatest care and skill in its adjustment; instead of such care and skill, we find more poor work here than anywhere. Fig. 7 is another illustration; here we have a club-tooth escape wheel, one of the teeth was broken and a piece of brass was soldered to the under side of the rim and the tooth filed to shape. It does not resemble the other teeth in size or form. I have known instances where teeth have been inserted so nicely it would bother any one to detect them, and would give good satisfaction, which illustrates the difference between a good and a poor workman.

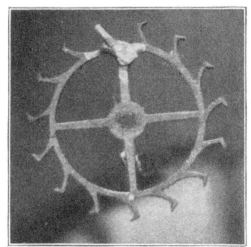

Fig. 7.

Fig. 8.

We do not often find a poorer piece of work than that shown in Fig. 8. We have a Swiss balance cock where the screws evidently did not hold in the regulator cap and the balance jewel was loose. These faults were corrected by

Fig. 9.

soldering on the cap and flowing solder around the jewel. The regulator was also soldered fast in the operation which rendered it useless, and there is no way of cleaning the jewels, and yet some one did such work and in all probability charged his customer for doing it.

Several pages were devoted to the mainspring in another article on that subject, but recently other specimens of very poor work have come to me, such as Fig. 9. It is hard to believe that any one could do such work, but it is a fact. From indications it would seem that the main spring would not hold on the arbor, so instead of putting in a new hook or dressing up the end of the spring properly, the inner end was soldered to the arbor as can be plainly seen, the spring soon broke.

Fig. 10 is equally as bad. It is hard to tell what was done in this case. The spring is much too thick, has only six

Fig. 10.

coils and nearly fills the barrel. It was too wide and the edges have been filed down, notice also the style of fastening on the outer end, a pin passing through the barrel and head and the end of the spring bent around it. How can any one do such work? They surely can have no conscience, but many such workmen seem to prosper while the more conscientious ones do not.

Perhaps none of the work is as poorly done as pivoting. We find some cases where the pivots are not in the center, where the shape is very poor, where the shoulders are not square and occasionally where pivots have been ground to a point. Several illustrations of the work will give us an ob-

Fig. 11.

Fig. 12.

ject lesson that should be effective. I trust none of my readers will recognize any of their work.

No work done by the ordinary watchmaker presents as great a variety of quality as that of pivoting, from this, one would judge the task is a very difficult one to perform. It does require considerable skill to replace a broken pivot in such a manner that it cannot be detected, but unless it is so done, the work is not satisfactory.

In Fig. 11 we have a photograph of a pinion that has been pivoted. The shoulder of the pivot is not square and no part of it is well polished. The temper was drawn and the color was never polished off.

233

Fig. 13.

Fig. 14.

234

WHY SOME WORKMEN FAIL TO SUCCEED.

At Fig. 12 is shown a pivoted staff. The pivot is not in the center, it is not cylindrical but is very much tapering, and the cone is badly formed, while the end which should be flat, comes nearly to a point.

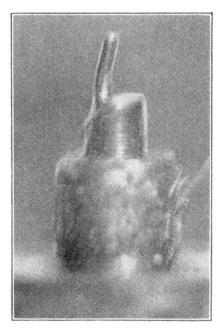

Fig. 15.

We have in Fig. 13 another pivoted staff. In this case, the cone is split, the plug does not fit properly and the general finish is not workmanlike.

There is no more common fault that causes us trouble than that shown in Fig. 14. The original pivot perhaps was long enough but not having the necessary end shake, was ground back until it was too short to pass through the balance jewel and reach the cap jewel; it must then bind on the cone; this is the cause of many watches acting poorly in one position and well in the others. All pivots having cap jewels should pass through the balance jewels far enough that the cap jewel will force it back enough to free the cone.

About as poor a specimen as I have seen is shown in Fig.

15. How any one could let such a piece of work leave his place of business is very hard to understand, but the results accomplished could be easily guessed. The pivot shown in Fig. 15 is very bad, but the staff illustrated in Fig. 16 would

Fig. 16.

go in about the same class. There are no square shoulders, no shape to the pivots, in fact, there is but little resemblance to a staff, yet even this was in use for some time. It may have been made by some beginner, even in that case he should have made a different use of his practice work.

Many workmen do not understand how to harden and temper a piece of steel wire so it will make a good pivot. They nearly always leave the steel too soft and it bends easily.

236

WHY SOME WORKMEN FAIL TO SUCCEED.

This is well shown in Fig. 17. The pivot was very small and the steel was quite soft, the watch evidently had a fall, the pivot did not break but was badly bent.

The pivot of the pallet arbor shown in Fig. 18 is about

Fig. 17.

as poor a specimen of square shoulder pivots as one often finds, but some one tried to get it to do the work of a pallet arbor.

Many good workmen while cleaning or repairing a watch,

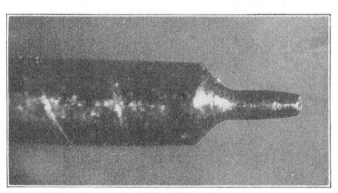

Fig. 18.

will take out the balance wheel or some of the other wheels after they have been oiled and lay them carefully on the bench paper, then as they put the movement together again, handle them again with the same care, and yet they have failed to observe one of the most important points. No matter how carefully the bench is cleaned and dusted, there are always particles of dust in the air and they soon settle on

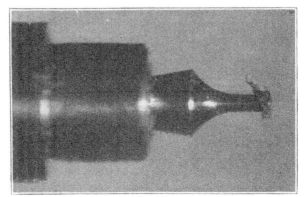

Fig. 19.

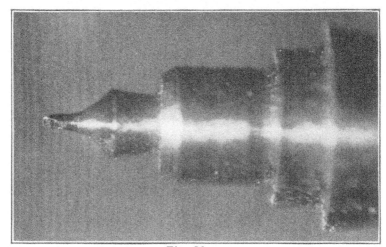

Fig. 20.

Fig. 21.

238

the paper; a pivot covered with oil will gather up much more of this dust than any one would imagine. Place such a pivot under the microscope and you will not be surprised that some pivots are soon worn nearly through. Fig. 19 shows the particles of dust on the end of a pivot that was perfectly clean before touching the bench paper. Unless such a pivot is always thoroughly cleaned with a piece of pith in order to remove this dust, it will be carried into the jewel by the pivot, and soon the pivot will begin to wear.

Many are very careless in polishing pivots and fail to keep them round. I have found many that were nearly flat on one or more sides like the one in Fig. 20. It is unfortunate that anyone would try to use a staff when so mutilated.

Some workmen seem to think pivots that are cap jeweled should have pointed pivots. They evidently get their idea from some of the alarm clocks. To illustrate, Fig. 21 shows the arbor of a pivoted detent used in a pocket chronometer. The whole escapement was cap jeweled and was a very nicely made movement. It was a disgrace to put such an arbor in a fine timepiece of this kind. After the new arbor was made, it performed its duty again as well as ever.

A peculiar method of fastening a jewel is show in Fig. 22. A graver has been used to dig up the plate around the jewel in order to make it hold, not a very elegant method.

There are many causes for a watch having a poor rate that are not so much the result of poor workmanship as carelessness, yet in one sense carelessness is poor workmanship. One may neglect to oil a pivot and in a short time it will begin to wear and cut until it is ruined and can be of service no longer. Fig. 23 shows the top pivot of a staff so badly worn that there is but little left of it. The side shake in the jewel would be excessive. Often in cleaning, a worn pivot is overlooked; every pivot should be examined carefully with a strong glass and polished when necessary and a new one put in when badly worn.

The pinion shown in Fig. 24 has the pivot badly worn. This was caused by not being well oiled; a pivot will not cut when properly oiled, but will run a great many years without perceptible wear, but the moment it gets dry, the wear begins.

Fig. 22.

Fig. 23.

WHY SOME WORKMEN FAIL TO SUCCEED.

A point that is often overlooked is the wear occasioned by the teeth of the fourth wheel acting against the leaves of the escape pinion, in some cases nearly one-half of the thickness of the leaves is worn away. This changes the depth between the two and the watch will stop and before one is

Fig. 24.

able to open the case, it is liable to start up and may run for hours before it will stop again. In some cases the fourth wheel can be either raised or lowered, so the teeth will act upon a part of the pinion that has not been worn and will perform as well as ever. If this can not be done, the only remedy will be a new pinion; a pinion that had to be replaced by a new one on account of being so badly worn is shown in Fig. 25.

We often find a brass jewel pin that poorly fits the notch in the lever, set in the roller, then we find one that is not set square and very often we find one that is much too small to fit the fork. These are very important things, as much of the motion is dependent upon this small part of the escapement. Often the jewel pin will be loose and escape our notice, this, too, will cause poor motion.

It will be impossible to enumerate all of the causes of poor motion in watches or to tell of all the poor work that

Fig. 25.

has been done that would give bad results; it would require a large book to tell them all, but enough has been shown to cause the conscientious workman to stop and think, then I am sure he will act, and in the future avoid doing poor work or allowing any to leave his bench that is not his very best. If every one would only do his best in everything he does, we would soon have a better quality of workmen and much less trouble with our work.

If the pages of this book have been the means of enlightening some of my fellow workmen and causing work that was formerly difficult for them to accomplish to be done in a quicker and better manner, if they are able to produce better work with the same effort as before, or if some thoughts expressed, will enable them to give better service to the public in general, the writer will feel fully repaid for the time and labor expended in writing them.

THE END.

Pocket Sun Dial.

Old Clepsydra or Water Clock.

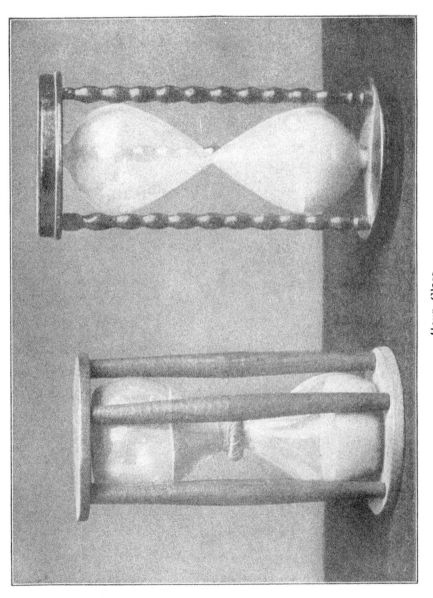

Hour Glass.

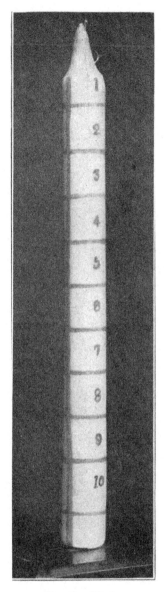

Candle Clock.

249

Lamp Clock.

INDEX

HOROLOGICAL DEPARTMENT

(School for Watchmakers)

Bradley
Polytechnic Institute

PEORIA, ILLINOIS

HOROLOGY HALL

Offers the most complete course in Watchwork, Jewelry, Engraving and Optics. Most efficient corps of instructors and finest equipment, at a cost most reasonable

For General Catalogue Address

HOROLOGICAL SCHOOL
PEORIA, ILL.

Other buildings contain extensive work in general education
Special catalogue

Printed in Great Britain
by Amazon

50054057R00149